KETO MEAL PREP GUIDE FOR BEGINNERS

The Complete Healthy Meal Prep Cookbook for Beginners to Lose Weight and Get Healthy

(Get Healthy and Feel Great on the Ketogenic Diet)

David McDougal

Published by Alex Howard

© David McDougal

All Rights Reserved

Keto Meal Prep Guide for Beginners: The Complete Healthy Meal Prep Cookbook for Beginners to Lose Weight and Get Healthy (Get Healthy and Feel Great on the Ketogenic Diet)

ISBN 978-1-990169-69-4

All rights reserved. No part of this guide may be reproduced in any form without permission in writing from the publisher except in the case of brief quotations embodied in critical articles or reviews.

Legal & Disclaimer

The information contained in this book is not designed to replace or take the place of any form of medicine or professional medical advice. The information in this book has been provided for educational and entertainment purposes only.

The information contained in this book has been compiled from sources deemed reliable, and it is accurate to the best of the Author's knowledge; however, the Author cannot guarantee its accuracy and validity and cannot be held liable for any errors or omissions. Changes are periodically made to this book. You must consult your doctor or get professional medical advice before using any of the suggested remedies, techniques, or information in this book.

Table of contents

Part 1 1
Introduction 2
Chapter 1: Breakfast Recipes 3
Egg and Pepper Bake 3
Meal Prep Oatmeal 3
Breakfast Smoothie 4
Ham and Feta Quiche 4
Spinach Artichoke Breakfast Bake 5
Oatmeal Bake 5
Zucchini Breakfast Burritos 7
Stuffed Breakfast Peppers 8
Simple Casserole 8
Beefed Up Yogurt 10
Breakfast Scrambler 10
Crockpot Casserole 11
Greek Casserole 11
Chapter 2: Lunch Recipes 13
Zucchini Pizza 13
Healthy Chicken Salad Sandwich 14
Copycat Salad 14
Quinoa Mixture 15
Salami Sammies 15
Fajita Bowls 16
Vegetable Soup 17
Cottage Cheese and Fruit 17

Just Like Instant Noodles 18

Grilled Tomato Cheese 18

Perfect Panini... 19

Protein Wraps... 20

Chickpea Salad... 20

Fried Chicken Bites 21

Holiday Sandwich 21

Chapter 3: Dinner Recipes 23

Beef of Mongolia.. 23

Spicy Southern Pot Roast.. 23

Honey Orange Chicken 24

Skinny Pasta Dish.. 24

Country Ham Casserole... 25

Skinny Tiki Masala 26

Taco Bake.. 26

Lighter Lasagna.. 27

Black and Bleu Burgers 28

Greek Kabobs... 28

Garlic Parmesan Chicken 29

Quick Chicken 29

Homemade Easy Helper 30

Honey Mustard Pork Chops 30

Skinny Mac and Cheese 31

Roasted Stuffed Peppers 31

Cheesy Rollups 32

MUSCLE BUILDING MEAL PLAN (7 DAYS).................................33

Day 1 – Breakfast: Poached Eggs with Wholemeal Flatbread 33
Day 1 – Lunch: Bean Salad with Bacon 34
Day 1 – Dinner: Risotto with Shrimp and Fennel 35
Day 2 – Breakfast: Baked Eggs with Mushrooms and Tomatoes 36
Day 2 – Lunch: Slow Cooker Beef and Cabbage 38
Day 2 – Dinner: Chicken and Goat Cheese 39
Day 3 – Breakfast: Frittata with Spinach and Pepper 40
Day 4 – Lunch: Pork Cutlets with Pepper 42
Day 4 – Dinner: Chicken Burritos with Quinoa 43
Day 5 – Breakfast: Yoghurt Porridge 44
Day 5 – Lunch: Steak and Tomatoes 45
Day 5 – Dinner: Congee with Soy Eggs 46
Day 6 – Breakfast: Protein Pancakes with Spinach 48
Day 6 – Lunch: Spanish Omelet with Chorizo 49
Day 6 – Dinner: Baked Salmon with Eggs 51
Day 7 – Breakfast: A Berry Omelet 52
Day 7 – Lunch: Moroccan Eggs with A Touch of Spice 53
Day 7 – Dinner: Squash Lasagna 54
WEIGHT LOSS MEAL PLAN (7 DAYS) 57
Day 1 – Breakfast: A Healthy Smoothie Bowl 57
Day 1 – Lunch: Broccoli in Squash Lasagna 58
Day 1 – Dinner: Roasted Cauliflower with Extra Flavor 60
Day 2 – Breakfast: Muffins from Pumpkin and Oats 61
Day 2 – Lunch: Carrot Soup 62
Day 2 – Dinner: Pear and Cottage Cheese Snack 64

Day 3 – Breakfast: Walnut Granola and Yoghurt Combo.........65
Day 3 – Lunch: Salmon with Ginger and Broccoli......................66
Day 3 – Dinner: Crispy Onions and Greens.................................67
Day 4 – Breakfast: Peanut Butter Scones with Chocolate Chips
....69
Day 4 – Lunch: Philly Steak Sandwiches......................................70
Day 4 – Dinner: Chicken Noodle Soup 72
Day 5 – Breakfast: Frittata with Broccoli....................................73
Day 5 – Lunch: Turkey Spring Rolls..74
Day 6 – Breakfast: Quick Banana Protein Shake.......................77
Day 6 – Lunch: Steak and Chutney 78
Day 6 – Dinner: Orange Beef 79
Day 7 – Breakfast: Apple Bars..81
Day 7 – Lunch: Rosemary Leg of Lamb 82
Day 7 – Dinner: A Quick Pizza...83
Conclusion..86
Part 2..87
Introduction 88
Eating Healthy and Staying Fit 90
Weight Loss Myth 91
How Weight loss Works...93
Prepping Yourself for Success-Tips for A Quick and Healthy Cooking...94
Nutrition Guidelines 99
Nutritional Instructions That Work...99
Understanding Calories 101
Calorie Count and Macronutrients 103

Meal Prep Made Easy – Adopt These Cooking Techniques . 105
Your Healthy Weekly Meal Plan 108
Meal Prep Week 1 109
Monday 109
Tuesday 109
Wednesday 109
Thursday 109
Friday 109
Saturday 110
Sunday 110
Meal Prep Week 2 110
Monday 110
Tuesday 110
Wednesday 110
Thursday 110
Friday 111
Saturday 111
Sunday 111
Meal Prep Week 3 111
Monday 111
Tuesday 111
Wednesday 111
Thursday 112
Friday 112
Saturday 112
Sunday 112
Meal Prep Week 4 112

Monday 112

Tuesday 113

Wednesday 113

Thursday 113

Friday 113

Saturday 113

Sunday 113

Shopping Guide and Food List 114

Fruits 114

Cereals & Grains 115

Flour 115

Nuts & Seeds 116

Drinks 116

Herbs & Spices 117

Meat, Fish. Eggs, Tofu & Legumes 117

Vegetables 117

Sauces & Condiments 119

100+ Healthy Meal Prep Recipes 120

Breakfast 121

Beef and Bacon "Rice" with Pine Nuts 121

Apple Pie Breakfast Farro 122

Chicken-Almond "Rice" 123

Raspberry Chia Breakfast Jars 124

Venetian "Rice" 125

Blackened Mexican Tofu, Greens, and Hash Browns 126

Smoky Bean and Tempeh Patties 127

Nuts and Seeds Breakfast Cookies 129

Japanese Fried "Rice"	130
Tilapia on a Nest of Vegetables	131
Balsamic-Mustard Chicken	132
Lonestar "Rice"	133
Lunch	134
Quick & Easy Tomato and Herb Gigantes Beans	134
Chinese Sticky Wings	135
Pappardelle with Cavolo Nero & Walnut Sauce	136
Wicked Wings	138
Veggie Sausage & Sun-Dried Tomato One Pot Pasta	138
Roast Chicken with Balsamic Vinegar	139
Orange-Five-Spice Roasted Chicken	140
Microwaved Fish and Asparagus with Tarragon Mustard Sauce	141
Orange-Tangerine Up-the- Butt Chicken	142
Almond-Stuffed Flounder Rolls with Orange Butter Sauce	144
Sherry-Mustard-Soy Marinated Chicken	146
Two-Cheese Tuna-Stuffed Mushrooms	147
Dinner	148
Cauliflower Rice Deluxe	148
Creamy Wild Mushroom One-Pot Gnocchi	149
Black Bean & Avocado Tacos	150
Spatchcocked or UnSpatchcocked Chicken with Vinegar Baste	151
Pan-Barbecued Sea Bass	152
Absolutely Classic Barbecued Chicken	153
Company Dinner "Rice"	154

Soy and Ginger Pecans 155
Hearty Quinoa Waffles 156
Salmon Stuffed with Lime, Cilantro, Anaheim Peppers, and Scallions 158
Turkey-Parmesan Stuffed Mushrooms 158
Saffron "Rice" 159
Sides 160
Cauliflower Puree 160
Chipotle-Cheese Fauxtatoes 161
Cheddar-Barbecue Fauxtatoes 161
Hobo Packet 162
Cauliflower Kugel 163
Little Mama's Side Dish 164
Gratin of Cauliflower and Turnips 165
Mushrooms in Sherry Cream 166
Avocado Cream Portobellos 167
Grilled Portobellos 169
Kolokythia Krokettes 170
Salads 171
Autumn Salad 171
Classic Spinach Salad 172
Spinach-Strawberry Salad 173
Summer Treat Spinach Salad 173
Dinner Salad Italiano 174
Chefs Salad 174
Vietnamese Salad 175
Cauliflower Avocado Salad 176

Sour Cream and Cuke Salad 177

Crunchy Snow Pea Salad 178

Parmesan Bean Salad 179

Fish and Seafood 180

The Simplest Fish 180

Ginger Mustard Fish 180

Aioli Fish Bake 181

Chinese Steamed Fish 181

Wine and Herb Tilapia Packets 182

Broiled Marinated Whiting 183

Whiting with Mexican Flavors 184

Brined, Jerked Red Snapper .... **Error! Bookmark not defined.**

Part 1

Introduction

Meal prep is beneficial because it takes the guesswork out of dieting.

You can benefit from meal prep when you do it the right way and when you have the best recipes possible.Keep in mind that the recipes are not created to be a diet but rather, to make it easier for you to prep your meals ahead of time.

Enjoy!

Chapter 1: Breakfast Recipes

Egg and Pepper Bake

8 - eggs
2 - green peppers
1 - chopped onion
1 - red pepper

Mix the eggs up into a large bowl and make sure that they are divided and the yolks are split.Chop up the peppers that you have and mix them in with the eggs; add the onions.Spray a baking dish with cooking spray and cook at350degrees for40minutes or until the eggs are firm in the middle.Divide into8pieces and put into containers

Meal Prep Oatmeal

2 - regular cup of oatmeal steel cut
½ - regular cup of coffee creamer, vanilla flavor
¼ - regular cup of diced peaches
1 - tsp of brown sugar

Divide all of the ingredients into two different parts.Put half of the oats into one small glass jelly tempered glass jar.Put the other half in a different one.Put the peaches on top of the oatmeal followed by the brown sugar.Pour the divided coffee creamer on top of the oats that are covered in peaches.Put the lid on and let sit at least overnight but up to 5 days in the refrigerator. Can be eaten cold while you are on the go or heat it up for 30 seconds in the microwave.

Breakfast Smoothie

1 - small bag of mixed berries
1 - small bag of spinach
1 - 16 oce. container vanilla yogurt
1 - tablesp. of honey
1 - tablesp. of chia seeds

Put all of your ingredients except for the chia seeds into a blender or a food processer and blend until they are well mixed up. There should not be any pieces of green from the spinach in there. Turn the blender off and add the chia seeds to the mixture. Divide into 6 parts and put into small glass jelly jars or some other type of container. Make sure that they are covered with a lid. You can store in the fridge for up to a week but make sure that you let them sit overnight so that they chia seeds have time to expand with the liquid of the mixture.

Ham and Feta Quiche

1 - large hamsteak
1 - 8 ouce. container of feta
1 - doz. Eggs

Cut the ham steak so that it is in bite sized pieces. Mix all of the eggs up together and make sure that there are no yolks that are stuck together. Add the ham and the feta to the eggs and stir well.Grease a baking dish and pour the mixture in.Set your oven to400and cook for about20-30minutes.The eggs should be firm.Divide it into6or8pieces and store in containers for up to a week.

Spinach Artichoke Breakfast Bake

1 - small bag of spinach
1 - can drained artichoke hearts
1 - dozen eggs
1 - teaspoo.Minced garlic
1 - splash EVOO

Put the olive oil into a pan and allow the spinach to cook until it is much smaller(do this after you wash the spinach).Mix the eggs up to ensure that there are no yolks that are still intact.Add the cooked spinach, the artichoke hearts and the garlic to the eggs.Spray a pie dish with cooking spray and set your oven to350.Cook the egg mixture for around25minutes or until the eggs begin to set up and the mixture no longer jiggles when you start to take it out of the oven.Cut into8different pieces and place in your containers.Serve with mustard or hollandaise sauce.

Oatmeal Bake

3 - regular cups of oatmeal that is steel cut
2 - bananas that are cut up into slices
1 - tsp baking soda
½ - tsp baking powder

1-16ouce.container of vanilla yogurt
1 - teaspoo.Of vanilla extract

Mix all of the ingredients except for the bananas up until they are well blended.Add the ingredients to a square baking dish and then top with the bananas.Put your oven on350degrees and allow the mixture to cook for about15minutes or until it a knife stuck in it comes out clean.Cut into4different pieces and save in your containers.

Banana Bread for Breakfast
1 - regular cup instant oats
1 - banana sliced up
1 - tsp cinnamon(or none if you use cinnamon applesauce)
1 - regular cup applesauce
1 - tsp baking powder
2 - eggs that are cracked

Mix all of the ingredients up and put them into a blender or a food processor to make sure that they are blended together and that they are able to be added together.Once you have blended them up, set your oven to350degrees and then grease a bread pan.Add the ingredients to the pan and cook for30minutes or until it is completely done.It should be slightly brown on the outside of it.Check it often to make sure that it is not cooking too fast and getting burnt while it is cooking.Cut into8slices.

Pumpkin Pie Oatmeal
3 - regular cup oatmeal that is not instant
1 - tin can of pumpkin
1 - tsp pumpkin pie spice
1 - cup vanilla yogurt
½ - regular cup almonds
Divide each of the ingredients into threes and then place the oats on the bottom of glass jelly jars.Mix the rest of the ingredients together until they are well blended up.Add the ingredients to the top of the oats.Cover with the jelly tempered glass jar lids and shake well until you are able to make sure that they are covered in the mixture.Allow them to sit in the fridge overnight so that they have time to soften the oats.They are able to stay in the fridge for one week.They are best eaten cold but you can also heat them up.Top with extra almonds if you desire.

Zucchini Breakfast Burritos
4 - tortillas
4 - eggs that are scrambled
¼ - regular cup cheddar cheese
¼ - regular cup sour cream
¼ - regular cup salsa
1 - zucchini chopped up and cooked until tender

Mix up the salsa, sour cream, and cheddar cheese.Spread it on each of the tortillas.Top with the zucchini and then the eggs.Roll it up like a burrito and put into plastic wrap.Heat up before you decide to eat it but make sure that you take the plastic wrap off first.Only heat to the point that the cheese begins to melt and the tortilla starts to soften up.

Stuffed Breakfast Peppers

4 - pepper that are cut in half(different colors are fun)
4 - eggs
1 - tblsp of olive oil
1 - tsp garlic powder
1 - tsp pepper
¼ - regular cup mozzarella cheese

Once the peppers are cut in half, line them up together in a square baking dish.Make sure that they are standing so that you are able to pour the ingredients into them.Brush them with olive oil.Mix the eggs with the pepper and the garlic powder and be sure that the yolks are all mixed up well.Pour them into the peppers and then top with the mozzarella cheese.Cook at400for20minutes.Broil the tops when done to make the cheese crispy.Allow them to cook and then wrap in foil and store in the fridge for up to5days.

Simple Casserole

1 - doz eggs
4 - roma tomatoes that are chopped up
1 - orange pepper that is chopped up
1 - tsp minced garlic
½ - regular cup cheddar cheese

Mix all of the ingredients up together and get more cheddar cheese if you want to put on top of the casserole.Grease a baking dish with cooking spray or with butter, put the ingredients into the dish and then cook for about40minutes on350or until the eggs are not longer jiggly.Be sure that you allow it to cook thoroughly.Top with the extra cheese.Make sure that it cools down before you cut into8pieces and put in your prep containers.

Breakfast Quesadilla
6 - tortillas, whole wheat is best
½ - doz eggs that are scrambled up
½ - regular cup salsa
1 - tsp chilli powder that you mix into the eggs
1 - oz pepperjack cheese

Place the tortillas into a pan and allow them to heat up.Put the eggs and the cheese on top.Put your salsa in a small container and then into your meal prep container.Add the other tortillas on top of the other ones.Heat up until the cheese melts.Cut into6pieces.Place6pieces from each into the container that you just put the salsa and store in the refrigerator for around one week.

Beefed Up Yogurt

1-16oz container of Greek yogurt
1 - tablesp of chia seeds
1 - regular cup of frozen fruit
½ regular cup - cottage cheese

Mix all of the ingredients together thoroughly but do not use a blender or even a mixer so that you do not disrupt the chia seeds.Make sure that you have four glass jars and then divide the mixture into each of the jars.You should be sure that there is enough room for it to expand because it will due to the content of the chia seeds.Allow it to sit for overnight so that the chia seeds get as large as possible.You can leave it in the refrigerator for up to one week as long as it is covered by the lid of the tempered glass jar.

Breakfast Scrambler

1 - doz eggs
1 - colby jack cheese you have shredded
1 - tblspoo milk
1 - onion that is chopped
1 - pepper that is chopped

Mix all of the ingredients up and put some olive oil into a very large frying pan.Allow it to heat up over medium high heat and then pour the ingredients into the pan.Cook until the eggs are fluffy and stir almost constantly to keep them from burning.Divide the mixture into6of your containers and make sure that it cools down before you put the lid on it.Serve with salsa or your favorite egg accompaniment.Do not store in the fridge for more than5days because the eggs will get rubbery.

Crockpot Casserole

1 - large bag of hashbrowns
1 - stick of butter
1 - doz of eggs
1 - hamsteak that is chopped up into pieces
4 - sausage links cut up into pieces
2 - peppers that are chopped
2 - tomatoes that are chopped

Coat the bottom of the slow cooker with the butter.Put the hashbrowns on the bottom and lay the rest of the stick of butter on top.Mix up the rest of the ingredients into a different container.Pour over the top of the hashbrowns.Cook on low for8hours or cook on high for four hours or until the eggs are thoroughly cooked and are fluffy.

Greek Casserole

1 - doz eggs
1-8oz of feta cheese
2 - roma tomatoes
1 - tablsp of minced garlic

1 - tesapoo of onion powder
1 - small tin can of black olive that are drained
Heat your oven up to400degrees.Use butter on a baking dish and make sure that it is coated the whole way.Mix the ingredients together and then pour them into the dish.Make sure that there are no yolks that are not broken up because they will not cook the whole way in the mixture.Once all of the ingredients are added to the mixture, cook for about40minutes.The mixture is done when the eggs are very firm even in the center and when you don't see it move at all while you are taking it out of the oven.

Chapter 2: Lunch Recipes

Egg Salad Pita

4 - pita pockets

8 - eggs---boiled

½ - regular cup mayo

¼ - regular cup mustard

¼ - regular cup relish

Cut your boiled eggs into small chunks.Mix them up with the mayo and mustard.Add the relish.Wrap your pita pockets in plastic wrap.Store everything in a container.Stuff pockets with egg salad just before eating.

Zucchini Pizza

6 - zucchini

36 - slices pepperoni

1 - regular cup mozzarella or pizza cheese blend

1 - regular cup pizza sauce(or pasta sauce)

Cut y our zucchini so that they are lengthwise and they look like they are boats.Scoop the seeds out of the zucchini but leave the flesh in place.Put your oven to400degrees and divide the pasta sauce between the slices of zucchini.Put three of the pepperonis on each of the "boats." Top with the cheese blend.Cook for10minutes or until they start to soften up a little bit.Cool down and divide into2zucchini per container.

Healthy Chicken Salad Sandwich

1 - tin can of chicken breast
½ - regular cup mayo
½ - regular cup Greek yogurt
1 - tblspoo of mustard
1 - teaspoo onion powder
½ - regular cup grapes, cut into quarters
¼ - regular cup almonds that are sliced up

Mix all of the ingredients together.Make sure that you are breaking up the chicken while you are mixing it so that it is chunkier in the chicken salad.Add all of the ingredients into one of your containers.Serve with whole wheat bread as a sandwich but make sure that you do not put the chicken salad onto the sandwich until just before you are ready to eat it for your lunch meal.

Copycat Salad

1 - large cardboard box of romaine lettuce
1 - pint of strawberries
1 - American pound of chicken that has been grilled
1 - small cardboard box of blueberries
1-8oz container of bleu cheese
¼ - regular cup of walnuts that are chopped up
¼ - regular cup of oats
2 - apples cut into small pieces

Cut your lettuce, your strawberries and your apples all at the same time.Mix all of the ingredients together and divide between6of the containers that you have for meal prep.Make sure that you do not put the dressing on until you are ready to eat but bleu cheese goes best with this salad.You can also use honey mustard or a simple vinaigrette for this salad.

Quinoa Mixture

1 - regular cup quinoa that has been cooked
1 - kiwi that has been cut up
10 - strawberrys that have been cut up
1 - tblsp of honey
1 - teaspoo of lime juice

After the quinoa has been cooked and has had the time to cool down, toss it in the honey and the lime juice.Add the fruit to it and mix it up so that the fruit is evenly distributed.Divide among three containers that you are using for meal prep.Serve with a salad or a sandwich.You can also double up the recipe and make enough that will work as a standalone lunch.Eat this cold which makes it great for a lunch to take to work.

Salami Sammies

8 - slices of pumpernickel bread
1 - container of cream cheese, softened
1 - tablesp of dill
1 - teaspoo of Worcestershire sauce

8 - slices of salami
1 - tablesp of chives

After the cream cheese has had time to soften up, mix it together with the dill and the chives making sure that they are evenly distributed.Once they are mixed together, add the sauce and stir it until it is thoroughly combined.Spread the mixture on each of the slices of bread.Top4of the slices with2slices of salami followed by the remaining4slices of bread so that you make4sandwiches.Wrap up and store in the fridge for up to one week or put into a meal prep container.

Fajita Bowls

1 - American pound cooked chicken breast
1 - can corn
1 - can black beans
3 - regular cup rice, cooked
1 - tempered glass jar of salsa

Make sure that you drain the black beans and rinse them off.Cut the chicken breast so that it is in bite sized chunks.In a large bowl, mix all of the ingredients together until they are thoroughly combined with each other.Make sure that they are going to be mixed up the right way.Top with cilantro or even with avocado if you are going to eat right away.Otherwise, simply divide into four meals and put them into your meal prep containers so that you can save them in the fridge for up to a week and you won't have to worry about them going bad.

Vegetable Soup

1 - large container of chicken broth
4 - potatoes peeled and cut into chunks
6 - stalks of celery
1 - large tin can of roasted tomatoes
1 - tin can of corn
1 - tin can of green beans
1 - tin can of pinto beans

Put all of the ingredients into a large stock pot and allow them to heat up over medium high heat until the potatoes begin to soften.Simmer for about three hours.When it is done, take it off of the heat and divide it into8of your meal prep containers.Serve with crusty bread or with any of the other lunches.Add some noodles to create a standalone meal that will stand *out* at lunch.

Cottage Cheese and Fruit

1 - cup of cottage cheese
1 - cup of fruit
1 - tablespoo of honey
1 - leaf of lettuce

Put the fruit and the honey into a blender and make sure that it is mixed up well enough to be able to form a jelly-like substance.Be sure that it is mixed up enough to eat and that you can enjoy it as much as possible.When you have done that, put the lettuce into one of your containers, top it with the cottage cheese and then the fruit mixture.Double, triple or quadruple the recipe to make several different lunches for your meal prep solution.

Just Like Instant Noodles

1 - cardboard box of noodles
1 - teaspoo of chicken boullion or1cube
1 - carrot chopped into small pieces
1-1onion chopped into small pieces or minced onion
1 - teaspo of parsley

Cook the noodles just until they start to get slightly soft but not until they are actually cooked up.Drain the noodles and put the rest of the ingredients into a bowl.Mix them up so that they are well combined and then divide them into4different glass tempered glass jar containers.Fill the containers with cold water until they are at the top of the tempered glass jar.Cover them with the lids and put the rings around them.Keep them in the fridge for anywhere from24hours to5days.When you are ready to eat them, simply remove the lid and heat up for about60seconds.

Grilled Tomato Cheese

4 - tomatoes that have been cut into small pieces

8 - slices of wheat bread
4 - slices of American cheese

Put your8slices of bread down onto a pan and top them with the cheese and the tomatoes.Allow it to heat up for just a few minutes or until the bread starts to get toasty.Put the empty slices on top of the full slices and flip over one time and cook for one minute.When done, wrap tightly in tinfoil and store in the containers that you have already set aside for meal prep.Serve within2days of creating so that they do not get too soggy.

Perfect Panini

8 - slices rye bread
8 - pieces of thick ham
4 - slices Havarti cheese
1 - tablesp of honey mustard
1 - red onion sliced up

Lay4of the slices of the bread on your panini press.Top it with the ham, the cheese, and the onions.Rub the honey mustard on the other four slices of bread.Put the bread down with the honey mustard touching the rest of the ingredients.Press down and make sure that they are not losing the ingredients.Close the panini press and make sure that it is fully cooked.The cheese should be melted well and the ingredients should be hot.Take them out and cover in foil then store in a container that you are using for meal prep.

Protein Wraps

2 - tortillas
2 - tablespo of peanut butter
2 - bananas sliced
1 - teaspo of honey

Mix the honey and the peanut butter up. Lay the tortillas out flat and then spread the peanut butter mixture on top of the tortillas so that they are coated. Lay the slices of banana on top of the peanut butter that is on the tortilla. Roll them up. You can leave them like this or you can cut them into slices so that they look like pinwheels. Store for up to three days in the fridge or a whole day out of the fridge. Be aware that they may get soggy because of the peanut butter.

Chickpea Salad

1 - tin can of drained and rinsed chick peas
1 - tin can of corn
1 - tin can of diced tomatoes
1 - regular cup of parmesan cheese
1 - tablespo of balsamic vinegar

Mix all of the ingredients together in a big bowl. Make sure that they are well combined and that they are properly mixed up. There should be no big chunks of parmesan cheese that are in there and the corn should have been drained when you first added it to the chickpeas.After it is all thoroughly combined, put it into6of the serving bowls that you have bought for meal prep.Make sure that you eat it within one week to keep it from spoiling because of the cheese that is in it.

Fried Chicken Bites

1 - American pound of breast of chicken
1 - regular cup parmesan cheese
1 - regular cup panko bread crumbs
2 - eggs
½ - regular cup pickle juice

Mix the egg and the juice up in a bowl.Cut the chicken up into small pieces.Dip it in the egg, the breadcrumbs, the cheese and the egg again and then lay on a pan.Put your oven on400and cook for30minutes until it begins to get crispy.When it is done, but your broiler on for one minute and allow it to get even crispier on the outside.Serve with a light sauce and place in one of your containers.This should be enough for4meals that are in the containers.

Holiday Sandwich

8 - slices of rye
4 - slices of turkey
4 - slices of cranberry sauce
¼ regular cup - gravy, premade

Take the gravy and rub it on 4 of the slices of rye. Top with the turkey and the cranberry sauce. Add the last four pieces of bread. Store in plastic wrap or put in the containers that you have set aside specifically for meal prep so that you can serve it. Save in the fridge for one week but try to eat sooner than that so that you do not have to deal with a soggy sandwich when you are trying to enjoy the flavors that come along with a traditional Thanksgiving. Heat up before you eat for the best flavor and a great hot meal right at your work desk.

Chapter 3: Dinner Recipes

Beef of Mongolia

1 - pound of skirt steak
1 - regular cup soy sauce
¼ - regular cup minced garlic
½ - regular cup orange sauce
1 - teaspoo of cornstarch

Mix the last four ingredients together in a bowl and make sure that they are well combined.Toss the steak in the mixture and then lay it into the bottom of a medium-sized slow cooker.Pour the rest of the mixture on top of the steak and put the lid on the slow cooker.Cook on low for six hours.When it is done, remove it from the slow cooker.Divide into four different containers and close the lids.Make sure that you eat it within a week of making it since it does have meat in it.

Spicy Southern Pot Roast

1 - large roast
1 - tempered glass jar of pepperoncini peppers
1 - stick of salted butter

Lay your roast into your large slow cooker.Pour the tempered glass jar of peppers over top of the roast.Cut the butter into six slices and place them on top of the roast, also.Put the lid on the slow cooker and cook for eight hours on low.Make sure that you do not lift the lid until after eight hours so that it can continue cooking the whole way and so that it will cook evenly.The roast can be divided into about10servings and put into your meal prep containers.

Honey Orange Chicken

3 - lbs of chicken breasts, cut into pieces that are bite sized
1 - regular cup of orange marmalade
1 - regular cup of honey
1 - regular cup of panko breadcrumbs

Put the honey, the marmalade and the breadcrumbs together into a large, gallon sized bag.Mix up the breadcrumbs with the other ingredients when they are in the bag.Put the chicken into the bag and make sure that it gets coated with the sticky mixture.Heat your oven so that it is at400degrees.Place the chicken evenly on a pan and make sure that the pieces are not overlapping each other.Cook for about20minutes and then flip over to the other side.Cook for another20minutes.The chicken should be165degrees on the inside and there should be no pink when it is done cooking. Divide into four different containers.

Skinny Pasta Dish

1 - cardboard box of whole wheat pasta

1 - tin can of diced tomatoes
1 - regular cup of parmesan cheese

Put the pasta into a pot of boiling water and allow it to cook until it is al dente.Make sure that you drain the pasta well and then return it back to the pot.Add the tomatoes to the pot and then allow it to cook for a few more minutes until it is completely heated again.Take it off of the heat, stir in the parmesan cheese.Divide into6different containers that you have made specifically for meal prep.Serve with a salad or any other side that can be used with your typical pasta dishes to make sure that you are going to be able to be full.

Country Ham Casserole

1 - hamsteak
½ - bag frozen peas, thawed out
2 - regular cup cheddar cheese
1 - cardboard box rotini noodles

Cook the noodles until they are al dente.Drain them and then add in the peas and the hamsteak.After, you can add in the cheese.Stir up so that they are all mixed together and put into a large baking dish.Heat your oven up to350degrees and then put the casserole into the oven for about10minutes.After that, take it out, sprinkle more of the cheddar cheese on it and put it back into the oven.Allow it to sit under the broiler for about1minute until the cheese melts and gets slightly crunchy on the outside.Serves eight servings that can be divided into meal prep containers.

Skinny Tiki Masala

1 - American pound chicken thighs
1 - tin can of diced tomatoes
1 - regular cup of heavy cream
1 - tablespo of garam masala
1 - tablespo of turmeric
1 - tablespo of garlic
1 - tablespo of ginger

Add the chicken, the tomatoes and the seasoning to a large pot.Allow it to cook over low to medium heat for about one hour or until the chicken is tender.Ramp the heat up to medium high and cook for another15minutes.The chicken should begin to fall apart.Take it completely off of the heat and allow to cool for a few minutes.Stir in the cream.Separate into eight different meal prep bowls.

Taco Bake

1 - American pound of ground beef, cooked
1 - snack size bag of cheesy tortilla chips
3 - regular cup taco cheese
1 - tempered glass jar of salsa
1 - regular cup of sour cream
1 - tin can of black beans drained and rinsed

After the ground beef has cooked, mix all of the ingredients up into a large bowl.Prepare a baking dish by spraying it with cooking spray.Pour the ingredients into the bowl and press down on them so that they are flat against the baking dish.Top with extra taco blend cheese and cook in the oven at350degrees for about25minutes or until it is hot throughout.Add any toppings that you want, divide it into six servings and put it in your meal prep containers.

Lighter Lasagna

1 - cardboard box of lasagna noodles
1 - container of light ricotta cheese
1 - container of diced tomatoes
1 - tempered glass jar of pasta sauce
2 - zucchini that have been diced
1 - eggplant that has been diced
1 - green pepper that has been diced

Make sure that you cook the lasagna noodles until they are done, according to the directions on the cardboard box.Lay the noodles down so that they form a bottom.Layer the ricotta, the sauce and all of the vegetables until you get to the last layer.End with the vegetables instead of one of the pasta layers so that you can make sure that you have the creamy top to it.Set your oven to400degrees and cook for30minutes or until the eggplant and the zucchini are completely done.

Black and Bleu Burgers

4 - hamburger patties
1-8oz container of bleu cheese
4 - brioche buns

Make sure that your patties are ready to be heated up.Turn your grill to about500degrees and make sure that it gets there.Put the burgers onto the grill and allow them to cook until the outside is slightly blackened but not until they are burnt.Put each of the burgers in a container and top with the bleu cheese.You can wrap up the buns separate so that they do not get soggy and put them into the container.Store in the fridge for up to three days.

Greek Kabobs

1 - sirloin steak or steak tips
1 - pint of cherry tomatoes
2 - zucchini chopped into small pieces
1 - tablespo of minced garlic

Rub the steak tips with the garlic and slide alternating with the steak tips, the tomatoes and the zucchini on kabob sticks.Make sure that you are alternating.Place the kabobs on the grill over500degrees and allow them to cook.Make sure that you let them cook slightly before because the tomatoes are very hot.Use tzatziki dip and place all of the ingredients together into a container.This should make about16kabobs so it is a good idea to divide them into fourths and put four of the kabobs into each of the containers.

Garlic Parmesan Chicken

1 - American pound of chicken breast that has no skin or bones
¼ - regular cup of parmesan cheese
1 - tablespo of minced garlic
1 - egg that has been beaten

Mix the parmesan and the garlic together.Put the chicken in the egg and then dip in the parmesan mixture.Put a metal cooling rack on the top of a pan and lay the chicken pieces on it.Cook in the oven at400degrees for40minutes or until the chicken is cooked the whole way through.This should make about four servings so divide them into your meal prep containers so that you can enjoy them for dinner.

Quick Chicken

1 - American pound of chicken breast that is cooked and cut up
1-8oz container of sour cream
1 - tin can of cream of chicken soup
1 - sleeve of Ritz crackers
1 - stick of butter

Mix up the sour cream and the chicken in a pot and put them over the low to medium heat setting on your stove.Allow them to come to a boil and then remove from the heat.Stir in the chicken to the ingredients.Pour the chicken into a baking dish that has been sprayed with cooking spray.Top the mixture with the Ritz crackers that have been crumbled up into pieces.Melt the butter and pour over top of the crackers.Turn your oven to350degrees and cook the mixture for25minutes.Divide into eighths.

Homemade Easy Helper

1 - American pound of ground beef
1 - cardboard box of rotini
3 - regular cup American cheese, shredded
½ - regular cup milk

Cook your ground beef until it is completely browned.Add the noodles and the milk to the ground beef in a large pan.Allow it to simmer for about10minutes or until the noodles are completely cooked.When they are, put the cheese into the mixture and stir up so that it is fully mixed together.Divide it into sixths and put it into your meal prep containers.Store in your fridge for up to one week.

Honey Mustard Pork Chops

1 - American pound of pork chops
1 - regular cup of honey mustard

1 - regular cup of panko

Put the honey mustard and the panko into a large gallon bag.Mix them up so that they are mixed completely and then make sure that they are divided evenly.Drop the pork chops in and then shake them up.Put your oven to400degrees.Lay your chops out on a pan and cook for20minutes.Take them out and flip them over then put them back in for another20minutes.Divide them and put them in your meal prep containers.

Skinny Mac and Cheese

1 - cardboard box of elbow macaroni
2 - tin cans of cheese soup
1 - regular cup of cheddar cheese
½ - regular cup milk

Put the noodles into a large pot and fill it with water.Heat on your stove with medium heat until the water is boiling.Allow to cook for about eight minutes or until the noodles are very tender.Add the cheese soup and the milk.Stir it up and allow it to cook for a few minutes but do not let it boil.Add the shredded cheddar cheese.Divide into eight bowls.

Roasted Stuffed Peppers

8 - peppers with the tops cut off
1 - American pound of ground beef that is cooked

1 - regular cup of pepper jack that is divided
1 - tin can of diced tomatoes
1 - onion that is chopped

Mix the tomatoes, the onion and half of the pepper jack together.Add them to the ground beef.Stuff the peppers with the mixture.Top the peppers with the rest of the pepper jack.Arrange them into a round pan and start your oven at350degrees.Cook for about20minutes or until the pepper is completely soft and easy to eat.

Cheesy Rollups

1 - American pound of chicken, cooked and chopped up
6 - tortillas
½ - bag of taco cheese

Mix up the cheese and the chicken.Put some of the mixture on each of the tortillas and roll them up.Lay them in a pan and allow them to heat up for about five minutes each.When they are done, take them out of the pan and then cut them so that they are pin wheeled into the way that they are supposed to look.Put about five of each of them into your meal prep containers.

MUSCLE BUILDING MEAL PLAN (7 DAYS)

Day 1 – Breakfast: Poached Eggs with Wholemeal Flatbread

Serves: 2 Preparation time: 10 minutes Cooking time: 6 minutes

This super easy egg recipe is a great start to the day for protein lovers! It also includes wholemeal flatbread for a dash of healthy carbs that will give you energy, and at the same time will also not interfere with your protein intake. You can add any other ingredients of your choice, but if you do, just make sure that they are both healthy and high in protein!

2 cups of chopped broccoli
1 cup of cherry tomatoes
4 eggs
2 whole meal flatbreads
1 tsp of oil
Salt and pepper to taste

- Bring water to the boil in a saucepan. Then cook your broccoli for about 10 minutes.
- Chop up the cherry tomatoes in halves. Set both aside for later.

- Bring a new saucepan of water to the boil again, and crack the two eggs inside the water to poach them. Cook for about 2 to 3 minutes.
- Heat oil in a pan and add the flatbreads to warm them up a little and give them some color.
- Serve everything on a dish and add salt and pepper to taste.

Per Serving: Calories: 383; Total Fat: 17g; Saturated Fat: 4g; Protein: 22g; Carbs: 31g; Fiber: 9g; Sugar: 4g

Day 1 – Lunch: Bean Salad with Bacon

Serves: 2 Preparation time: 20 minutes Cooking time: 20 minutes

Not every lunch has to be heavy, especially if you have a long day ahead of you. A bean salad is not only a great source of protein, it is also very easy to digest and won't deplete your energy levels. You can use any beans in your recipe, but white beans have a distinct and creamy flavor which many people find absolutely delicious. Plus, the crunchy bacon ads a whole new level of taste!

4 slices of bacon
1 can of beans
3 tbsp of vinegar

3 tbsp of olive oil

3 tbsp of mustard

1 tbsp of chopped chives

Salt and pepper to taste

- Fry the bacon in a large pan until it is very crispy.
- When done, crumble the crispy bacon so that it can later be used as garnish on top.
- Mix the beans, vinegar, oil, mustard and seasonings until they are all covered.
- Refrigerate for two hours if you'd like the salad to be cold.
- Garnish with bacon and chives on top.

Per Serving: Calories: 250; Total Fat: 9g; Saturated Fat: 2g; Protein:32g; Carbs: 14g; Fiber: 9g; Sugar: 2g

Day 1 – Dinner: Risotto with Shrimp and Fennel

Serves: 2 Preparation time: 50 minutes Cooking time: 45 minutes

Seafood is a great dinner choice when you are taking care of your health because it is very easy on the stomach. Shrimp especially are great to give your food a rich layer of taste, without making the whole thing super heavy. The rice is creamy and delicious, and the fennel adds freshness to the whole dish.

4 tbsp of butter
1 chopped fennel
1 chopped onion
2 cups of rice
¾ cup of white wine
Salt and pepper to taste
8 cups of chicken broth
1 pound of shrimp
1 ounce of parmesan

- Melt the butter into a large pot. Add the onions and fennel and cook until soft. About 12 minutes.
- Add all of the rice and stir it in.
- When the rice is combined, add the white wine and seasoning. Cook until the wine has evaporated.
- Next add the broth, slowly, and cook until everything is combined. It will take about 25 minutes for the risotto to combine properly.
- Add the shrimp and cook it for a couple of minutes.
- Add parmesan as garnish when ready to serve.

Per Serving: Calories: 320; Total Fat: 12; Saturated Fat: 4; Protein:28g; Carbs: 12g; Fiber: 6g; Sugar:4g

Day 2 – Breakfast: Baked Eggs with Mushrooms and Tomatoes

Serves: 2 Preparation time: 15 minutes Cooking time: 30 minutes

This unique recipe makes a lovely baked dish as your source of protein and vitamins for breakfast. It takes very little time to prepare, and is a great dish for the whole family to enjoy. The saucy texture is filling and has enough energy to last for the whole day. You can add extra bacon on top if you like for more flavor, but for this particular recipe we will exclude it.

1 cup of mushrooms
1 garlic clove
2 thyme leaves
2 tomatoes
2 eggs
1 cup of rocket

- Preheat oven to 375°F.
- Place the mushrooms bottom side up onto a baking dish or baking tray. Top with thyme on top, cover with foil and then bake for 20 minutes.
- When mushrooms are done, remove the foil and add halved tomatoes on the side.
- Carefully break two eggs into each mushroom cup.
- Bake for another 10 minutes.
- Cool slightly before serving.

Per Serving: Calories: 240; Total Fat: 13g; Saturated Fat: 5g; Protein:22g; Carbs: 8g; Fiber: 4g; Sugar:5g

Day 2 – Lunch: Slow Cooker Beef and Cabbage

Serves: 2 Preparation time: 20 minutes Cooking time: 8 hours

This is great dish for a slow cooker, and one that will be packed full of delicious ingredients. The meat makes it high in protein, while the cabbage adds both taste and fiber, and is of course great for your digestive system. The beef becomes so tender that it is almost effortless to break it to pieces! This is a delicious, healthy lunch for all protein lovers to enjoy.

4 thyme leaves
1 tsp of caraway seeds
3 pounds of beef brisket
1 pound of carrots
½ of a green cabbage
1 pound of potatoes

- Combine the beef with the thyme and the seeds into your slow cooker.

- Add roughly chopped carrots, potatoes and cabbage with 1 cup of water (or follow slow cooker manufacturer instructions for water level).
- Cook for 8 hours on a low setting on your slow cooker.
- Cool before serving.

Per Serving: Calories: 320; Total Fat: 17; Saturated Fat: 7; Protein: 28g; Carbs: 12g; Fiber: 3g; Sugar:5g

Day 2 – Dinner: Chicken and Goat Cheese

Serves: 2 Preparation time: 20 minutes Cooking time: 30 minutes

Chicken is a favorite dinner ingredient because it is affordable, easy to cook, and delicious! It is also high in protein and can be combined with almost any ingredient in your fridge. For this recipe, we will use the creaminess of goat cheese to add complex flavors that will soon become your favorite dinner.

½ cup of olive oil
½ cup of chopped parsley
½ tsp of crushed pepper
2 ounces of goat cheese
4 ounces of chicken breasts

- In a bowl, combine goat cheese, pepper, half of the olive oil, and the parsley.
- Heat the oil in a pan and cook each chicken breast until done. Usually 6 minutes per side.
- Let the chicken cool a little before serving so that it doesn't melt the goat's cheese.
- Add the creamy mixture on top when ready to serve.

Per Serving: Calories: 240; Total Fat: 12g; Saturated Fat: 3g; Protein: 31g; Carbs: 11g; Fiber: 4g; Sugar:3g

Day 3 – Breakfast: Frittata with Spinach and Pepper

Serves: 2 Preparation time: 10 minutes Cooking time: 40 minutes

Even though this may not be the kind of breakfast that you would make in a rush, the good thing about it is that you can easily make it the day before. Not only is this recipe high in protein because of the eggs, but it is also high in vitamins and minerals because of the delicious peppers. This recipe is a guide, but you can add any vegetables of your choice to add additional crunch.

3 eggs
2 cups of cottage cheese
1 clove of garlic
½ cup of grated parmesan
2 cups of spinach
1 tsp of nutmeg
1 cup of cherry tomatoes

- Preheat the oven to 170°F. Line a baking tin with parchment paper.
- Beat the eggs in a bowl until they are fluffy.
- Add half of the parmesan, cottage cheese, garlic, spinach, and nutmeg.
- Pour the mixture into the baking tin, and top with tomatoes and the rest of the parmesan.
- Bake for 40 minutes, or until golden.
- Cool a little before serving.

Per Serving: Calories: 198; Total Fat: 4g; Saturated Fat: 3g; Protein: 22g; Carbs: 5g; Fiber: 4g; Sugar:4g

Day 4 – Lunch: Pork Cutlets with Pepper

Serves: 2 Preparation time: 20 minutes Cooking time: 25 minutes

Ho ho, talk about a hearty meal! This recipe is not only high in protein, it is also a great dish for lunch gatherings when the whole family wants to tuck in! Sometimes, it can be difficult to develop the perfect dish on a high protein diet and to avoid things that are always repetitive, but this dish is both original and will help to give you a change of taste.

3 tbsp of olive oil
4 pork cutlets
Salt and pepper to taste
2 bell peppers
2 shallots
½ cup of pitted olives
½ cup of parsley leaves
1 tbsp of vinegar

- Heat half of the oil into a pan.
- Season the pork and cook it on both sides until brown and cooked to taste.
- Heat the other half of the oil in a different pan, and cook the peppers and shallots.

- Cook until all the ingredients soften.
- Add the olives, parsley and vinegar to the skillet near the end of the cook for a couple of minutes.
- Serve the pork with the vegetables.

Per Serving: Calories: 270; Total Fat: 7g; Saturated Fat: 3g; Protein: 30g; Carbs: 8g; Fiber: 2g; Sugar:6g

Day 4 – Dinner: Chicken Burritos with Quinoa

Serves: 2 Preparation time: 25 minutes Cooking time: 25 minutes

We all need some time off after a long day at work every now and then, and what better way to end your day than with burritos? Except, these burritos have been developed to be both healthy and high in protein, so you don't have to feel any regret when eating them! This is another recipe that is fun to share with others.

2 whole wheat tortillas
1 cup of shredded, cooked chicken
1 cup of black beans
1 cup of cooked quinoa
1 cup of grated mature cheese
1 cup of cilantro
½ cup of Greek yoghurt
1 avocado

1 cup of salsa

- Heat up a pan on medium heat.
- Warm each tortilla for about a minutes on each side.
- Add the salsa by spreading it on the bottom of the tortilla.
- Next, add some shredded chicken on top with a layer of beans.
- Now add a layer of quinoa followed by a layer of cheese.
- Top with yoghurt and cilantro.

Per Serving: Calories: 180; Total Fat: 9g; Saturated Fat: 5g; Protein: 28g; Carbs: 11g; Fiber: 4g; Sugar:3g

Day 5 – Breakfast: Yoghurt Porridge

Serves: 2 Preparation time: 5 minutes Cooking time: 5 minutes

Not every morning needs to start with a super boost of protein. Sometimes, it is more important to think carefully about your gut and make sure that your digestive system is ready to take on the day! Yoghurt is a fantastic ingredient to start the day, and can be mixed with a variety of different ingredients if needed.

We like to serve it with mangoes, but you can choose another topping if you wish.

3 tbsp of porridge
2 cups of yoghurt
1 cup of mango (or fruit of choice)

- Add 1 cup of water into a non-stick pot. Bring the water to a simmer.
- Add the oats to the water and gently stir until the oats are cooked.
- Wait until the porridge thickens before removing it from the heat.
- Let the porridge cool before serving.
- Add mango as a topping (or another fruit).

Per Serving: Calories: 184; Total Fat: 2g; Saturated Fat: 0g; Protein: 13g; Carbs: 11g; Fiber: 4g; Sugar:8g

Day 5 – Lunch: Steak and Tomatoes

Serves: 2 Preparation time: 15 minutes Cooking time: 25 minutes

Steak is a good meal to have in the middle of the day because it is high in protein but also because it takes longer to digest than other sources of protein. A good steak is very good for your well-being, but it really is important that you eat it for lunch and not for dinner. This recipe is very saucy, so it's great people who live tomato sauce or gravy.

2 tbsp of olive oil
2 pieces of steak
Salt and pepper to taste
1 cup of tomatoes
¼ cup of fresh oregano

- Heat half of the oil in a large skillet, bringing the heat to high.
- Season the steaks with salt and pepper, and then cook them on both sides. This will usually take about 6 minutes per side.
- Let the steaks rest for 10 minutes before slicing.
- Heat the remaining oil in a new skillet.
- Add the tomatoes and peppers and cook until soft.
- Serve the rested steaks with the vegetables on top.

Per Serving: Calories: 305; Total Fat: 7g; Saturated Fat: 1g; Protein: 21g; Carbs: 7g; Fiber: 4g; Sugar:2g

Day 5 – Dinner: Congee with Soy Eggs

Serves: 2 Preparation time: 10 minutes Cooking time: 25 minutes

A little bit of Asia to your daily meals! Asian dishes are known to be highly protein-based and also full of as many nutrients as can possibly be packed into a recipe. Congee is also known as a gentle dish which provides a slow release of energy when needed. This is why it is also good as an evening meal, because it will satisfy your hunger while also being light enough to digest.

2 cups of rice
2 cups of chicken stock
1 cup of cooked, shredded chicken
1 slice of ginger
2 spring onions
¼ cup of roasted peanuts
13 cup of coriander
3 eggs
1 cup of soy sauce
1 tbsp of sugar

½ cup of water

- Start by making the delicious homemade soy eggs.
- Bring a pan of water to the boil, and then boil the eggs for about 6 minutes. When the eggs are cooked, cool and peel them.
- In a small bowl, mix the soy sauce with the sugar until the sugar dissolves. Add ½ cup of water and leave aside for two hours.
- To make the congee, combine the rice with the chicken stock and bring them to a boil.

- Cook for about 25 minutes until you get a soupy texture. If needed, you can add small amount of water.
- When done, top the congee with the soy eggs, ginger, spring onions, peanuts, and coriander.

Per Serving: Calories: 210; Total Fat: 9g; Saturated Fat: 3g; Protein: 18g; Carbs: 9g; Fiber: 6g; Sugar:4g

Day 6 – Breakfast: Protein Pancakes with Spinach

Serves: 2 Preparation time: 15 minutes Cooking time: 25 minutes

Finally, pancakes! They are a lovely start to a morning and will guarantee to brighten up anyone's day. But if you are on a high protein diet, you need the pancakes to help you with nutrition as well. This recipe channels a little bit of Popeye, because it adds spinach to increase the flavor and the nutrients! It's very easy to make and you can even make a few in advance for dinner.

1 ½ cups of butter milk
1 egg
2 cups of spinach
1 cup of buckwheat flour

½ tsp of paprika
2 tbsp of oil

- Prepare a pot of water to boil for the spinach.
- Beat the eggs, and then combine them with the butte milk in a blender.
- Add the spinach to the boiling water for just a minute or two until it wilts.
- Add to the process with the eggs and the butter milk and process it too.
- Combine all of the dry ingredients and gradually mix them with the puree. Make sure that the whole thing is very smooth.
- Heat oil in a frying pan and then cook the pancakes as you normally would.

Per Serving: Calories: 241; Total Fat: 14g; Saturated Fat: 3g; Protein: 17g; Carbs: 10g; Fiber: 4g; Sugar:1g

Day 6 – Lunch: Spanish Omelet with Chorizo

Serves: 2 Preparation time: 15 minutes Cooking time: 15 minutes

This is a very special treat for people who love Chorizo! It is a very easy lunch to make, but the great thing about it is that it is both high in protein and also very easy to take with you to work. It will stay fresh for a few hours, and it can also be easily reheated if needed. You can always top it with extra vegetables or a bit of creaminess if you like.

2 tbsp of olive oil
1 onion
2 ounces of Chorizo
¾ pound of potatoes
Salt and pepper to taste
½ cup of parsley
5 large eggs
1 cup of shredded Cheddar cheese
1 red onion

- Heat the oven to 400° F.
- Add half of the oil to a skillet and heat to a high temperature.
- Add the onion and cook for about 5 minutes.
- Add the chorizo, potatoes and seasoning and cook until soft.
- Add the parsley and the beaten eggs. Stir very gently to make sure that eggs have spread out a little bit.
- Cook for another 15 minutes in the oven.

- Cool before serving.

Per Serving: Calories: 245; Total Fat: 10g; Saturated Fat: 3g; Protein: 22g; Carbs: 11g; Fiber: 6g; Sugar:6g

Day 6 – Dinner: Baked Salmon with Eggs

Serves: 2   Preparation time: 5 minutes   Cooking time: 15 minutes

Salmon is a great source of healthy fats and protein. Because it's a fish, it is the perfect ingredient for a dinner. Toped with the eggs it really brings a boost to your daily protein needs, and it is also a great recipe if you need to top up your daily protein levels a little bit. It is best to keep this recipe as simple as possible, because you really don't want to eat anything too heavy for dinner.

2 whole wheat bread rolls
½ cup of butter
2 slices of smoked salmon
2 eggs
½ cup of chives

- Heat the oven to 350° F.

- Slice the top off of each roll, and remove some of the bread in the middle to make room for the salmon and the eggs.
- Arrange the rolls onto a baking sheet.
- Coat the inside of each roll with melted butter.
- Place a slice of salmon into each roll, and then crack an egg on top.
- Bake for 15 minutes.
- Cool before serving.

Per Serving: Calories: 220; Total Fat: 11g; Saturated Fat: 3g; Protein: 18g; Carbs: 11g; Fiber: 4g; Sugar:3g

Day 7 – Breakfast: A Berry Omelet

Serves: 2 Preparation time: 10 minutes Cooking time: 5 minutes

Not every breakfast needs to be savory just because you are on a high protein diet! Although fruit is not the key ingredient of a high protein diet, fruit has a lot of nutritious properties which can help your body with energy levels and health. This easy recipe is a quick, on the go omelet with a sweet touch. You can add any fruits of your choice to the top.

1 egg
1 tbsp of skimmed milk

½ tsp of cinnamon
1 tsp of oil
1 cup of cottage cheese
1 cup of chopped strawberries (or any fruit)

- Beat the egg into a bowl until it becomes fluffy.
- Add the milk and combine with the egg.
- Next, combine the mixture with cinnamon.
- Heat a frying pan with the oil.
- Cook as you would normally cook an omelet.
- Let it cool a little before adding the fruit.

Per Serving: Calories: 264; Total Fat: 12g; Saturated Fat: 5g; Protein: 21g; Carbs: 18g; Fiber: 4g; Sugar:16g

Day 7 – Lunch: Moroccan Eggs with A Touch of Spice

Serves: 2 Preparation time: 10 minutes Cooking time: 20 minutes

Spicy food is delicious for lunch, especially if lunch is with friends! This easy meal is full of flavor and great with its high protein levels. It is a truly hearty meal, so it is best enjoyed slowly. Bring a little taste of Morocco to your home. You can also add a slice of whole wheat bread to help with the leftover sauce if you like!

2 tbsp of oil
1 onion
2 cloves of garlic
1 tsp of coriander
1 cup of vegetable stock
2 cans of chickpeas
2 cans of crushed tomatoes
2 courgettes
1 cup of baby spinach
4 eggs

- Heat the oil in a large pan.
- Fry the onion and the garlic until soft. Add the stock and the chickpeas and continue to cook.
- Cover and simmer for 5 minutes.
- Mash about one third of the chickpeas, and then add the tomatoes. Cook for another 10 minutes.
- Add the courgettes and cook for another 5 minutes, and then finally add the spinach and cook for one more minute.
- Break the eggs into the mixture and cook them whole for about 2 minutes.
- Cool before serving.

Per Serving: Calories: 242; Total Fat: 10g; Saturated Fat: 2g; Protein: 16g; Carbs: 22g; Fiber: 8g; Sugar:7g

Day 7 – Dinner: Squash Lasagna

Serves: 2 Preparation time: 15 minutes Cooking time: 4 hours

Lasagna may not be the first thing that comes to your mind when you are deciding on dinner, however, this lasagna is made with the help of a slow cooker! The squash makes it very healthy and adds another source of protein to the dish. Because this dish is made in a slow cooker, it is very easy to prepare and can be made for about 3 days in advance so that you don't have to prepare dinner every night.

10 ounces of squash puree
½ tsp of ground nutmeg
1 cup of ricotta
1 ounce of baby spinach
Salt and pepper to taste
12 lasagna noodles
3 cups of mozzarella

- Combine the nutmeg and the squash in a bowl.
- In a different bowl, combine the spinach with the ricotta cheese and the seasoning.
- At the bottom of your slow cooker, layer about ½ cup of the squash. Add a layer of lasagna noodles and then a layer of the ricotta cheese.
- Continue this process until all ingredients have been used up.
- Cook in the slow cooker for 4 hours.

Per Serving: Calories: 280; Total Fat: 12g; Saturated Fat: 4g; Protein: 30g; Carbs: 13g; Fiber: 4g; Sugar:3g

WEIGHT LOSS MEAL PLAN (7 DAYS)
Day 1 – Breakfast: A Healthy Smoothie Bowl

Serves: 1 Preparation time: 10 minutes Cooking time: 10 minutes

Smoothie bowls can be easily tailored to suit your favorite tastes and textures. When it comes to weight loss, a smoothie bowl is full of nutrients without having to worry about too many calories. This bowl uses chai seeds and berries for delicious flavor, and we've added some delicious topping for extra flavor white still keeping it low in calories.

1 banana
1 cup of mixed berries
½ cup of soy milk
½ cup of pineapple chunks
1 kiwi
1 tbsp of sliced almonds
1 tbsp of coconut flakes
1 tsp of chia seeds

- In your blender, combine the soy milk, banana and berries. Blend until the mixture is completely smooth.

- Add the chia seeds and let them sit inside the smoothie for at least 15 minutes.
- Top the smoothie bowl with pineapple, almonds, and coconut.

Per Serving: Calories: 328; Total Fat: 10g; Saturated Fat: 3g; Protein: 9g; Carbs: 40g; Fiber: 12g; Sugar: 16g

Day 1 – Lunch: Broccoli in Squash Lasagna

Serves: 2 Preparation time: 20 minutes Cooking time: 40 minutes

No reason why you can't eat lasagna just because you're watching your weight! The essence of lasagna isn't in the pasta, it's in the creamy textures that come inside this delicious dish. Luckily, there are healthy and low-calorie ways to develop these textures. This recipe uses broccoli, peppers, and mozzarella to bring you the taste of lasagna that you'll love!

3 pounds of spaghetti squash
1 tbsp of olive oil
1 bunch of broccolini
3 cloves of garlic
1 tsp of red pepper (crushed)

2 tbsp of water
1 cup of shredded mozzarella
½ cup of parmesan cheese
½ tsp of salt
½ tsp of ground pepper

- Preheat the oven to 450°F and make sure that the racks are inside the oven.
- In a microwave dish, add the squash with a little bit of water. Microwave for about two minutes. Or until soft.
- Add oil to a skillet and bring it to a high heat.
- Add the broccolini, garlic, and red pepper and cook until it softens. Add a little water to make sure that the broccolini has enough liquid to cook in.
- Remove the squash from the microwave. Create spaghetti-like strings with the help of a fork.
- Take the shells of the squash and add them to a baking pan that can stand the heat of a broiler.
- Mix the squash fresh with mozzarella, seasoning and half of the parmesan. Add the remaining parmesan on top.
- Bake in the oven for about 15 minutes, and then an extra 2 minutes under the broiler to become crispy.

Per Serving: Calories: 190; Total Fat: 11g; Saturated Fat: 5g; Protein: 11g; Carbs: 14g; Fiber: 6g; Sugar: 2g

Day 1 – Dinner: Roasted Cauliflower with Extra Flavor

Serves: 4 Preparation time: 15 minutes Cooking time: 30 minutes

Crispy, roasted cauliflower with a topping of melted parmesan is a delicious evening meal. Very easy to make, simple ingredients, and you can share it with the whole family without worrying about calories! Take care with the amount of cheese though, because even a little too much parmesan can add extra calories that you don't want.

8 cups of cauliflower
2 tbsp of olive oil
½ tsp of salt
Pepper to taste
2 tbsp of balsamic vinegar
½ cup of parmesan cheese

- Preheat the oven to 450°F.
- Add oil and seasoning to the cauliflower and toss it until all of the ingredients are covered.
- Spread the cauliflower onto a baking sheet and roast it for about 15 minutes.

- Toss the cauliflower in vinegar and then sprinkle it with cheese. Return it to the oven to roast for about 10 more minutes.
- Cool a little before serving.

Per Serving: Calories: 150; Total Fat: 10g; Saturated Fat: 2g; Protein: 7g; Carbs: 9g; Fiber: 3g; Sugar: 5g

Day 2 – Breakfast: Muffins from Pumpkin and Oats

Serves: 21 Preparation time: 20 minutes Cooking time: 40 minutes

Muffins are an easy to make breakfast that is especially good if you are one of those people who has a lot of things to do on the go and doesn't have enough time in the morning to prepare breakfast. The great thing about this recipe is that you can prepare it days in advance and then not have to worry about it for the whole week!

2 cups of oats
1 tsp of baking powder
1 tsp of pumpkin spice
13 tsp of baking soda
½ tsp of salt

2 eggs
1 cup of pumpkin puree
½ cup of brown sugar
3 tbsp of grapeseed oil
1 tbsp of vanilla extract

- Preheat oven to 350°F. Coat a muffin tin with cooking spray and set to the side.
- Finely grind the oats in a blender until they have a smooth texture.
- Add salt, pumpkin spice, baking powder, and baking powder and mix together.
- Add eggs, pumpkin puree, brown sugar, oil, and vanilla, and puree again until everything is smooth.
- Fill the muffin cups with the mixture and bake for about 20 minutes.
- Cool before eating.

Per Serving: Calories: 78; Total Fat: g; Saturated Fat: 1g; Protein: 2g; Carbs: 13g; Fiber: 5g; Sugar: 2g

Day 2 – Lunch: Carrot Soup

Serves: 8 Preparation time: 20 minutes Cooking time: 40 minutes

Carrot soup is full of Vitamin E, which is essential for your skin and your eyes. In fact, carrots are one of the few ingredients which actually releases more nutrients when it is heating than when it is raw. This is why it is now the perfect time to make it! And, it serves as a delicious, light lunch too.

1 tbsp of butter
1 tbsp of olive oil
1 onion
1 stalk of celery
2 garlic cloves
1 tsp of chopped parsley
5 cups of carrots (chopped)
2 cups of water
4 cups of chicken broth
Salt and pepper to taste

- In a pot, heat the oil and the butter until the butter melts completely.
- Add the onion and the celery and cook until both become soft.
- Add garlic and parsley and continue to cook.
- Add the carrots and the broth, stir, and then cover the pot and cook for 25 minutes.
- Puree the whole thing when done.

Per Serving: Calories: 160; Total Fat:9g; Saturated Fat: 3g; Protein: 7g; Carbs: 7g; Fiber: 5g; Sugar: 4g

Day 2 – Dinner: Pear and Cottage Cheese Snack

Serves: 1 Preparation time: 5 minutes Cooking time: 0 minutes

Pears and cottage cheese are a great combination! As a dinner meal, it is a great choice for when you have had enough food during the day and you just want to end the day with something light. You can exchange the pears with another fruit like apples if they are not in season.

1 pear
¼ cup of cottage cheese
1 tbsp of pepitas

- Cut the fresh pear into slices.
- Add cottage cheese into a cup and to with the pepitas.
- Dip the pears into the cheese as you eat.

Per Serving: Calories: 130; Total Fat: 2g; Saturated Fat: 1g; Protein: 8g; Carbs: 28g; Fiber: 6g; Sugar: 11g

Day 3 – Breakfast: Walnut Granola and Yoghurt Combo

Serves: 20 Preparation time: 10 minutes Cooking time: 45 minutes

For a good breakfast, you need a good balance between nutrients, fibers, and energy. This recipe is very convenient because one granola prep is enough for 20 servings! This means you will have almost an entire month of healthy breakfasts just by cooking once. The yoghurt will keep everything refreshing, but remember to always choose low fat yoghurt.

Cooking spray
2 cups of oats
1 cup of bran cereal
½ cup of puffed wheat cereal
½ cup of chopped walnuts
½ cup of sugar-free syrup
2 tbsp of canola oil
13 tsp of cinnamon
5 cups of low-fat yoghurt

- Preheat oven to 325°F. lightly coat a baking pan with non-stick spray.
- In a bowl, combine the cereal, oats, and walnuts.

- In a small bowl combine the syrup and the cinnamon. Then add it to the dry cereal mix and combine until everything is equally coated.
- Bake in the oven for about 30 minutes, or until golden brown.
- Serve with fresh low-fat yoghurt.

Per Serving: Calories: 230; Total Fat: 9g; Saturated Fat: 1g; Protein: 13g; Carbs: 4g; Fiber: 6g; Sugar: 8g

Day 3 – Lunch: Salmon with Ginger and Broccoli

Serves: 4 Preparation time: 20 minutes Cooking time: 25 min

Salmon is always a great ingredient, full of healthy fish oils and healthy salt minerals. This recipe uses ginger to add a special layer of flavor to a fish that so many people just eat plain. The broccoli is there for extra fiber and Vitamin C. very affordable and endlessly delicious!

2 tbsp of sesame oil
2 tbsp of tamari
1 tbsp of vinegar
1 tbsp of ginger
½ tsp of salt

8 cups of broccoli
1 tbsp of molasses
1 ½ pounds of salmon
2 tsp of toasted sesame seeds

- Preheat oven to 425°F. Prepare a baking sheet onto a baking tray and set it aside.
- Whisk half of the salt with vinegar, tamari, ginger and oil. Then add the broccoli and coat it with the mixture.
- Transfer the broccoli onto the baking tray.
- Mix the molasses into what remains of the broccoli marinade.
- Roast the broccoli for about 10 minutes.
- In a different pan, coat the salmon in the molasses marinade and fry it in a hot pan for about 6 minutes on each side.
- Sprinkle everything with sesame seeds.

Per Serving: Calories: 310; Total Fat: 12g; Saturated Fat: 3g; Protein: 36g; Carbs: 23g; Fiber: 3g; Sugar: 5g

Day 3 – Dinner: Crispy Onions and Greens

Serves: 2 Preparation time: 30 minutes Cooking time: 45 minutes

This is an innovative dinner recipe that will end your day with a final boost of healthy ingredients. The dandelion leaves are full of antioxidants, which will feed your body with a high amount of vitamins before bed time. It's very cheap to make and it is also easy to digest. The crispy texture makes it a wonderful light snack-dinner as you unwind for the night.

5 cups of dandelion greens (or kale if you can't find them)
1 tbs of olive oil
2 onions
2 garlic cloves
1 bunch of cilantro (finely chopped)
Juice of 1 lemon
½ tsp of salt

- Bring a pot to the boil and cook the dandelions leaves until they soften up.
- When it is done, place the drained dandelion leaves on a kitchen towel and squeeze until everything is completely dry.
- Heat up oil into a pot and cook the dandelions until they become golden and crispy.
- Add the onion and the garlic and cook those until crispy too.
- Remove from the heat and add salt, lemon juice, and cilantro.

Per Serving: Calories: 90; Total Fat: 6g; Saturated Fat: 1g; Protein: 2g; Carbs: 7g; Fiber: 6g; Sugar: 6g

Day 4 – Breakfast: Peanut Butter Scones with Chocolate Chips

Serves: 16   Preparation time: 20 minutes   Cooking time: 40 minutes

Did you ever think that scones could be both delicious and low in calories? These are the perfect scones for people who are lowering their weight and they are a great treat in the morning. The yummy peanut butter will help satisfy your morning cravings, and the chocolate is a dash of something extra special.

2 cups of flour
½ cup of whole wheat flour
1 tbsp of baking powder
½ tsp of salt
13 cup of peanut butter (organic)
3 tbsp of butter
½ cup of brown sugar
½ cup of milk
2 eggs
½ cup of chocolate chips

- Preheat oven to 400°F. lightly grease a baking tray and set it aside.
- In a bowl, combine the flours, salt, sugar and baking powder. Add butter and peanut butter, and combine until you get a crumbly texture.
- In another bowl, whisk the eggs and the milk until fluffy. Then slowly add them to the dry ingredients until everything is smoothly combined. Add chocolate chips and mix those in as well.
- Create a circle shape from the dough and cut it into 16 wedges.
- Bake for about 16 minutes or until golden brown.
- Cool before serving.

Per Serving: Calories: 180; Total Fat: 7g; Saturated Fat: 2g; Protein: 5g; Carbs:25g; Fiber: 2g; Sugar: 9g

Day 4 – Lunch: Philly Steak Sandwiches

Serves: 4 Preparation time: 15 minutes Cooking time: 40 minutes

This high-protein, low-calorie sandwich is a great treat for when you are having a hard time with your diet and you really want something to add a little zest in your food life. This sandwich favorite has been simplified a little bit in this recipe to get rid of all of the unhealthy ingredients, but it is still the delicious sandwich that everyone loves!

1 ounce of beef steak
½ tsp of garlic pepper
Cooking spray
2 sweet peppers
1 onion
4 whole-wheat buns
½ cup of shredded Cheddar

- Preheat the broiler.
- Trim all of the fat from the steak to make sure that it is lean.
- Place the steak onto a rack and broil it for about 20 minutes.
- Coat a skillet with cooking spray, and add the onion and the pepper (both chopped). Cook for about 10 minutes until tender.
- Place the buns on a baking tray and broil them for about 5 minutes.
- Assemble the sandwich with the sliced steak, and onion and peppers on top.

Per Serving: Calories: 300; Total Fat: 12g; Saturated Fat: 5g; Protein: 25g; Carbs: 26g; Fiber: 4g; Sugar: 8g

Day 4 – Dinner: Chicken Noodle Soup

Serves: 6 Preparation time: 35 minutes Cooking time: 3 hours

This recipe is very light! It is soup that you can easily eat as a light dinner, and you can even take it with you to work the following day because it is easy to store. The ingredients are affordable and easily available, and the soup itself is of course delicious!

Cooking spray
1 pound of chicken thighs
4 cups of chicken broth
½ cup of leeks
½ cup of celeriac
½ cup of carrots
13 tsp of salt
½ tsp of black pepper
½ cup of sour cream
1 tbsp of flour
2 tbsp of lemon juice
1 tbsp of dill
1 cup of egg noodles

- Coat a skillet with the cooking spray.
- Add the chicken and cook until light brown.
- Remove any excess fat.

- Transfer the chicken to a slow cooker, and add all of the vegetables.
- Cover and cook on a high setting for about 3 hours.
- Combine the sour cream and flour at the end and add it to the soup.
- Add the noodles at the end and cook for 10 more minutes.

Per Serving: Calories: 190 Total Fat: 5g; Saturated Fat: 1g; Protein: 20g; Carbs: 17g; Fiber: 9g; Sugar: 4g

Day 5 – Breakfast: Frittata with Broccoli

Serves: 2 Preparation time: 15 minutes Cooking time: 20 minutes

This is a great breakfast that's full of both protein and fiber. It is very quick to make and has the nutrient elements to keep your energy stable throughout the day you can add any other vegetables of your choice, but make sure that you pay attention to the calories so that you don't go overboard.

2 eggs
2 egg whites
1 tbsp of low-fat milk
½ tsp of black pepper

1 tsp of olive oil
½ cup of broccoli
½ cup of onion
1 cup of bread cubes
3 tbsp of cheddar cheese
¼ cup of parsley

- Preheat oven to 375°F.
- Whisk milk, eggs and pepper into a bowl.
- Pour into a heated skillet on a medium heat.
- Add onion and broccoli on top.
- When the frittata is cooked, add cheese and bread cubes on top. Sprinkle with parsley.

Per Serving: Calories: 250, Total Fat: 12g; Saturated Fat: 4g; Protein: 16g; Carbs: 20g; Fiber: 3g; Sugar: 3g

Day 5 – Lunch: Turkey Spring Rolls

Serves: 1 Preparation time: 10 minutes Cooking time: 15 minutes

This super light lunch is great both at home and as a lunch that you can take with you. The healthy ingredients make it a light meal that will let you focus on your work throughout the day, and won't cause any insulin spikes. If you like, you can add a layer of spice to the dish with some chili flakes.

1 tbsp of vinegar
½ tbsp of Sriracha sauce
½ cup of coleslaw mix
13 cup of cilantro
2 sheets of rice paper
2 ounces of turkey
½ cup of thinly sliced cucumber

- For dipping, mix Sriracha sauce with the vinegar.
- Combine the cilantro and coleslaw mix in another bowl.
- Fill a bowl with warm water, and dip each sheet of rice paper for about a minute until it softens a little.
- Boil the turkey for about 15 minutes. Drain when done.
- To assemble, place half of the turkey on each rice sheet, add vegetables on top and top with the sauce before you roll them.

Per Serving: Calories: 190; Total Fat: 1g; Saturated Fat: 0g; Protein: 14g; Carbs: 27g; Fiber: 3g; Sugar: 3g

Day 5 – Dinner: Avocado and Bean Wrap

Serves: 4   Preparation time: 20 minutes   Cooking time: 25 minutes

Time to make healthy dinners fun again! This delicious avocado and bean wrap is low in calories but very high in protein and fiber. It is very quick to make and will be more than filling for a light dinner. It is also a great dinner idea for friends and family that are coming over.

2 tbsp of vinegar
1 tbsp of oil
2 tsp of chipotle chili
13 tsp of salt
2 cups of cabbage
1 carrot
13 cup of cilantro
1 can of white beans
1 avocado
½ cup of cheddar cheese
4 whole-wheat wraps

- Whisk the salt, chipotle chili, vinegar and oil in a bowl.
- Add cilantro, carrot and cabbage, and toss them in the mixture.
- Mash the avocado with the beans in another bowl to create a creamy spread.
- To assemble the wrap, add the avocado and bean cream on the bottom. Top it with the carrot and cabbage mix and then add cheese on top.

Per Serving: Calories: 320; Total Fat: 16g; Saturated Fat: 3g; Protein: 12g; Carbs: 30g; Fiber: 5g; Sugar: 2g

Day 6 – Breakfast: Quick Banana Protein Shake

Serves: 1 Preparation time: 10 minutes Cooking time: 5 minutes

Start your day with a healthy protein shake with very few calories! This quick recipe is great when you don't have any time to literally make anything else. This shake tastes exactly like a chocolate shake, but without all the unhealthy ingredients of course!

1 banana (frozen)
½ cup of red lentils (cooked)
½ cup of non-fat milk
2 tbsp of cocoa powder
1 tsp of maple syrup

- Combine all the ingredients together in a blender and blend until everything is completely smooth.
- For best taste, leave in the fridge for 10 minutes to chill.

Per Serving: Calories: 290; Total Fat: 2g; Saturated Fat: 1g; Protein: 15g; Carbs: 27g; Fiber: 9g; Sugar: 4g

Day 6 – Lunch: Steak and Chutney

Serves: 6 Preparation time: 20 minutes Cooking time: 2 hours

This lunch takes some time to make, but it is well worth it! The steak becomes very tender and breaks apart as soon as you touch it. The chutney gives it a bit of vibrance and introduces a healthy fruit alternative to the usual sauces that go with steaks

1 pound of steak
1 cup of pineapple juice
½ cup of mango chutney
1 tbsp of pineapple juice
1 tbsp of vinegar
1 clove of garlic
13 tsp of salt
1 tsp of cornstarch

- Prep the steak by removing all of the unhealthy fat. Then make 1 inch slices along the steak for visual effect.
- To prepare the marinade, combine the salt, garlic, vinegar, rum, chutney and juice. Place the marinade into a plastic bag and then add the steak inside of the plastic bag. Leave the steak to marinade between 2 and 24 hours.

- When the steak is done, grill it for about 20 minutes (depending on how you like your steak).
- In a different pot, add the remaining marinade, add the cornstarch and cook until it thickens.
- Slice the steak and add the sauce on top to serve.

Per Serving: Calories: 230; Total Fat: 7g; Saturated Fat: 3g; Protein: 25g; Carbs: 15g; Fiber: 1g; Sugar: 2g

Day 6 – Dinner: Orange Beef

Serves: 4 Preparation time: 20 minutes Cooking time: 25 minutes

This beef recipe is delicious and full of ingredients that are both affordable and easy to find. This is also a great recipe for dinner parties when you want to make sure that you are still on track with your diet and that everything that you are eating is still low in calories.

8 ounces of green beans
2 tsp of sesame seeds
½ cup of orange juice
2 tbsp of soy sauce
1 tbsp of sesame oil
1 tsp of cornstarch
1 tsp of orange peel

Cooking spray
1 cup of green onions
1 tbsp of grated ginger
2 cloves of garlic
1 tsp of oil
12 ounces of steak
2 cups of brown rice
2 oranges (peeled and sectioned)

- Start by cooking the green beans for about 6 minutes in a sauce pan.
- In a skillet, cook the sesame seeds for a few minutes until roasted and until you can start to smell their fragrance.
- For the sauce, combine the orange juice, sesame oil, soy sauce, orange peel and corn starch.
- Coat a skillet with the cooking spray and fry the ginger, onions, and garlic. When those are done, add the beans and toss together.
- Add the beef and cook them all together to absorb the flavors.
- Add orange sections as a topping.

Per Serving: Calories: 320; Total Fat: 10g; Saturated Fat: 2g; Protein: 21g; Carbs: 30g; Fiber: 6g; Sugar: 3g

Day 7 – Breakfast: Apple Bars

Serves: 16 Preparation time: 20 minutes Cooking time: 1 hour

Add a little sweetness to your breakfast without the guilt of calories! These apple bars taste just like apple pie but have far less sugar. One preparation makes about 16 bars, which can be eaten at room temperature or chilled. You can even freeze them for up to six months!

1 cup of flour
1 cup of oats
½ cup of sugar
½ tsp of baking powder
¼ tsp of salt
13 cup of vegetable oil
1 egg
5 apples
2 tbsp of lemon juice
3 tbsp of brown sugar
2 tbsp of flour
1 tsp of cinnamon
½ tsp of ginger

- Preheat oven to 350°F.

- In a bowl, combine the oats, 1 cup of flour, salt, baking powder and sugar. Add the oil and the egg and mix until you get a crumbly texture.
- Add half of the mixture into a baking pan.
- In another bowl, combine the sliced apples with ginger, the remaining flour, cinnamon, brown sugar and lemon juice.
- Layer the apple over the crust and then add the remaining crust on top of the apples.
- Bake for 50 minutes.

Per Serving: Calories: 160; Total Fat: 5g; Saturated Fat: 1g; Protein: 2g; Carbs: 30g; Fiber: 4g; Sugar: 7g

Day 7 – Lunch: Rosemary Leg of Lamb

Serves: 15 Preparation time: 30 minutes Cooking time: 2 hours

This is quite the feast for a lunch! But it well-worth the time that it takes to make. This absolutely delicious meal is very hearty and yet low in calories because it is mainly protein-based. It is also great because it comes out with many servings, which you can then spread out for the rest of the week.

5 pounds of leg of lamb

2 tsp of rosemary
1 tsp of salt
1 tsp of black pepper
3 cloves of garlic
1 cup of water

- Preheat the oven to 375°F.
- Trim all the fat from the lamb.
- Mix the salt, rosemary and pepper in a bowl. Rub this mixture evenly all over the meat.
- Add the garlic in small cuts into the leg of lamb.
- Roast for about 2 hours or until done.

Per Serving: Calories: 140; Total Fat: 4g; Saturated Fat: 1g; Protein: 23g; Carbs: 0g; Fiber: 0g; Sugar: 1g

Day 7 – Dinner: A Quick Pizza

Serves: 2 Preparation time: 10 minutes Cooking time: 15 minutes

Sometimes, you really do need a bit of pizza to make the end of a day much better than it is. This quick and healthy mini pizza recipe will help to soothe any cravings that you may be having for this food. It is also low in calories and will not interfere with your diet at all! Invite your friends over and have a feast!

1 whole-wheat flatbread
3 tbsp of ricotta cheese
½ cup of dried tomatoes
3 tbsp of crumbled feta cheese
Cooking spray
2 eggs
½ cup of arugula
2 tsp of balsamic glaze
Salt and pepper to taste

- Preheat oven to 450°F.
- Place the flatbread onto a baking tray and bale for about 3 minutes.
- Add a layer of ricotta cheese first and then add the dried tomatoes on top. Then add the feta cheese and the parmesan as well.
- Bake for another 5 minutes or until the cheese melts.
- Fry the egg in a skillet. Cook the egg to however you like it (both hard and soft boiled are ok).
- In another bowl, combine the seasoning with the balsamic glaze and the arugula.
- Add the arugula onto the pizza and then the egg on top of that.

Per Serving: Calories: 212; Total Fat: 11g; Saturated Fat:5g; Protein: 17g; Carbs: 19g; Fiber: 2g; Sugar: 6g

Conclusion

Thank you for reading my cookbook!I hope that you enjoyed the recipes and they can help you toward your health journey.

If you liked the book, please feel free to let me know in a review!

Part 2

Introduction

Many people are addicted to new communities today, the free market quickly, cheaply and easily. The entire marketing campaign is designed through these three weaknesses of current human status.

When entering a world of diet control to lose a lot of weight, there are many obstacles to success. Most of these are mental obstacles. The problem is that the person whose goal is to lose weight is to get carried away for a while they need to break a habit (or maybe 20 habits) to reach the right mental state to achieve their goals.

One of these mental obstacles is the adoption of new meals and a new general way of eating. Someone who is severely overweight usually has a history of eating highly processed foods or eating in restaurants regularly.

Highly processed foods have tons of preservatives, sugars, flour, and other things that occupy a high range of blood glucose index. This means that someone who eats like this regularly completely crushed the Yielding of the pancreas in the long term. The pancreas is a fantastic member, but it can only be a lot over time.

Restaurants are selling their food in large sectors. Unfortunately, it seems that humans innately planned to store and prepare for a thin load. Therefore, because modern societies have been ruined by comfort, the meshing of large portions and the instinct to store has contributed significantly to an obese nation.

That said, food preparation helps you prepare meals that are healthy, easy, and incredibly effective when they help you lose fat, reverse the effects of type 2 diabetes, and increase overall immune system health and Yield.

Maybe, you're wondering if the preparation for the meal applies to you and what it is to her...

Not prepared for food is often considered something you do when you want to lose weight, or save time. But there are a few reasons why everyone should do this. Here are some additional tips on why someone should consider doing it, at least for some time:

You save money: One of the main reasons to start food preparation is to save you money. That's because you're able to buy more food in bulk. Consider how much you store with purchase meat and vegetables in size, rather than just purchasing small portions you need for one or two meals. Then you can prepare your meals, and get all the other benefits as well. Also, you will save money by making many different meals and avoiding eating.

This allows you more time during the week: If you are someone who often skips home-cooked meals cooked during the week, because you don't have much time to cause other work and responsibilities, preparing the meal will be right for you. Choose a day of night or weekend when you have some extra time, and prepare or cook lots of meals for weeks. This way, all you have to do is make your meals together and a little partial heating or cook the rest of the week.

You can eat healthily: Preparing food ensures that you eat healthy foods every meal is carefully planned. You will build some healthy foods at once, often using fresh or frozen production, lean protein, and other natural ingredients. It also helps you learn the control section. Use the preparation containers to prepare meals containing compartments that separate the different parts of the meals with the appropriate part size.

Preparing your meals in advance is not difficult to make. Start by accepting the fact that it's a little time, especially the first time you do it if you've never tried it before. There may be weeks when you want to include individual meals.

Eating Healthy and Staying Fit

Eating healthy and staying fit is the best thing you can do for your body and health. Proper exercise and proper amounts of nutrients can help your organization achieve this health and keep it up to date. Also, there may be health products you can buy to help you along the way.

People often do not know what to do to fit or eat healthily. They don't know what they are doing wrong. But really, anyone needs little guidance in their lives.

There are ways that you can exercise fitness that is right for you, your body type, and your personal goals. You can also find a diet that will give you the right amount of nutrients and keep your body feeling and look healthy. Maybe you have tried everything, and you don't get the results you're looking for. Perhaps you will have about it for the blindfold and don't know where to seek out some answers to your health questions, but searching the internet will give you many options of information.

Weight Loss Myth

The myth of weight loss has been around for many years, whether they appear on the Internet or in any popular magazine, a hot topic among people of faces and people has been aware of. This article will help you get great detail of 6 weight loss myths that people think they will lose weight. These myths are prevalent and are often sought by people who plan to lose weight. Believing that this myth can do you more harm than good. Discover the answers behind the most prevalent weight loss myth.

- ☐ Dramatically reduce calorie reduction in weight loss faster: When you drastically reduce your caloric intake, you send your body to "hunger mode." In this case your body starts saving calories and using less energy to store them for future use. Your metabolism is reduced, and you tend to gain weight rather than losing it.

- ☐ Strictly chase after a diet, leave no room to cheat: when you eliminate some food groups or focus on only one of them, you are required to get less effective results. Soon you feel tired of eating the allowed food group and may feel truly private, and this increases your chances of leaving a hard diet and docking in the group of fatty and Forbidden Foods, leading to weight gain.
- ☐ Eating late at night can increase my weight: this is one of the most famous legends commonly sought by bodybuilders and the conscious form of people. It doesn't matter if you are eating on a day or midnight, your body converts all extra calories into fat in a period. Eating a light snack before bed prefers to surprise you to sleep better.
- ☐ Get rid of your favorite food: Treat your favorite food from time to time to help you control your calorie intake and prevent you from choking. Moderation is the perfect way to find success in your weight loss mission.
- ☐ You should eat between meals: Well, many of us may follow this weight loss myth. Conversely, a light snack between meals will help control your blood sugar levels, improve your metabolism and help you burn more calories. This also enables you to manage your calorie intake while eating great meals.
- ☐ Fat is not good: most people who struggle to lose weight often follow these weight loss myths. Fat is also an essential part of our diet. Fats help to improve the flavor and aroma of food. Some good

fatty acids that will help you achieve your weight loss program Omega3 fatty acids found in crustaceans and fishes.

So, if you're looking for any of the weight loss myths mentioned above, then I'm sure the answer behind them will help you change your appearance.

How Weight loss Works

Metabolism is a process that explains how your body converts food to energy. Most of your calories are burned as a result of your resting metabolic rate. Your Resting Metabolic Rate is responsible for essential Yields such as standing, thinking, sitting, food digestion, etc. Your metabolic rate of exercise measures the energy costs that result from exercising. This can range from the physical activity of light to extreme fitness routines. If you want to lose weight you have to burn more calories than you take, and you will have to find ways to increase your metabolic rate to rest and exercise.

☐ Get started by eating less. The goal is not to starve; the goal is to get the balance. Eat your food in smaller quantities often. Generous proportions will give you the energy you need. If you feel full, stop eating. If you are hungry again, eat again. If you eat four or five meals a day instead of three standards, you are better off to control your weight.

☐ Many of our bad eating habits come from unprepared. If you suddenly become severely hungry, and in a ready position, the solution you find to your hunger dilemma is rarely a healthy option. Healthy decisions in life are deliberately made--- arrangements before the time of your choice. If you extend your meals throughout the day, you will have to plan. If you have a healthy eating option with you or have a ready when you are home, then you can avoid the need for lousy emergency feeding options.

☐ You have to exercise almost every day. Find ways to make the physical activity fun. Increasing your metabolic rate of exercise should be something you would expect. Think about the benefits. If you can burn more calories than you ingest, you will lose weight. Use makes sense on many levels. Just find a way to make it fun, and you will enjoy the results

Eating habits and healthy exercise are the basis for any substantial weight loss program. But the conclusion of your long-term weight management strategy is entirely up to you.

Prepping Yourself for Success-Tips for A Quick and Healthy Cooking

You are changing your lifestyle and want to eat healthier, but your obstacle is keeping the snare is the preparation of food.

"It takes too long! Is there a way to create "healthy eating" fast and easy?
of course!

You can find a friend who likes to cook every meal for you... Or you can hire a personal chef. But if this option isn't for you now, here's sure of the tips that can make your time in the kitchen "fast" and, perhaps, more enjoyable...

- ☐ Make your preparations more efficient. There are two main ways to do this:
- Select a specific day and time to make necessary preparations for the week (Sunday or afternoon, for example)
- Get ready to begin food preparation immediately after returning from your grocery shopping trip.

I always prefer the second option: food preparation immediately after purchase. Separate the meat into separate parts before storing it in a refrigerator or freezer. Wash vegetables and fruits and save them. If you are going to consume them in a short time, cutgrindjulianadice and set them aside, so you don't have to when you start cooking. (Note: Always prepare this food according to keeping food as fresh as possible.)

- ☐ Food planning in advance. The cooking time is often increased because you spend time trying to figure out what to eat in the first place! If you plan your meals (i.e. honestly, also your purchase speed), you can speed up your cooking time. The reward for this is that, after a few weeks, you will have an excellent Rolodex instruction and choose through rotation.

- [ ] The night before the start... Or in the morning. Drag the frozen meat part of the freezer to start melting. Organize the ingredients you need, either together in the refrigerator or at the top of the counter, whichever is appropriate. When possible, do the essential preparation of food in the form of cutting or tightening. Remove the kitchen appliances you need to bypass. Now you're ready to roll.
- [ ] Prepare your kitchen. Have you made all the tools you wantyour kitchen needs to be active and efficient? If yes, continue immediately (!) until cooking. If not, time to buy some kitchen appliances!
- [ ] Raw eat. Apart from revealing that "eating raw" saves cooking time because, well, there isn't, taking a lot of organically perched vegetables and fruits (and some meats) raw, or close to, actually benefiting your body. The more "live" the food, the more accessible and useful it is for your body. (Think: Life is supported by life. Dead, or too much significantly baked support ...)
- [ ] Examine your current schedule. Is there anything you can't stand" that you can remove from your program to create more free time? Get them out! Can you rearrange your schedule to create more time to do what you want? Reorganizes! Of course, if you have a lot of time and "something you can't stand" is your kitchen and preparation, see 7 and 8 below... Or consult with me and we'll understand

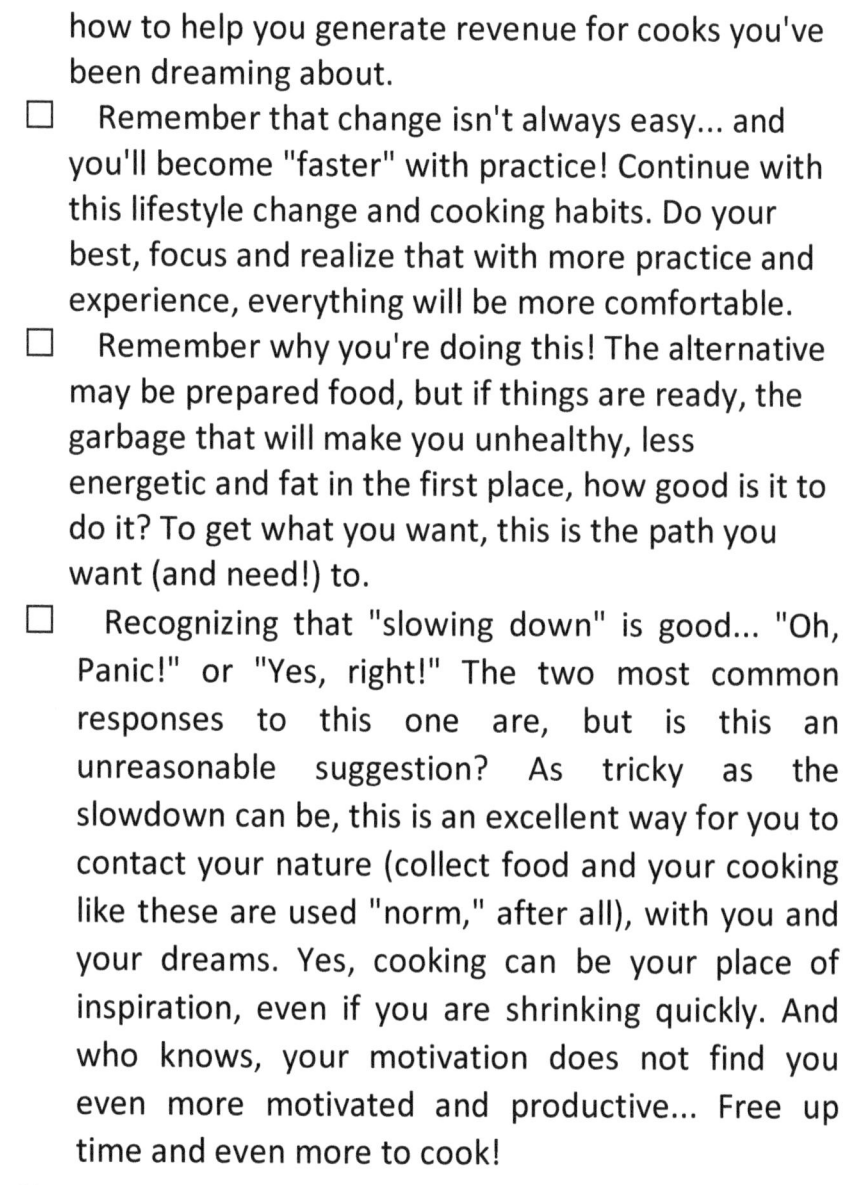

how to help you generate revenue for cooks you've been dreaming about.
- ☐ Remember that change isn't always easy... and you'll become "faster" with practice! Continue with this lifestyle change and cooking habits. Do your best, focus and realize that with more practice and experience, everything will be more comfortable.
- ☐ Remember why you're doing this! The alternative may be prepared food, but if things are ready, the garbage that will make you unhealthy, less energetic and fat in the first place, how good is it to do it? To get what you want, this is the path you want (and need!) to.
- ☐ Recognizing that "slowing down" is good... "Oh, Panic!" or "Yes, right!" The two most common responses to this one are, but is this an unreasonable suggestion? As tricky as the slowdown can be, this is an excellent way for you to contact your nature (collect food and your cooking like these are used "norm," after all), with you and your dreams. Yes, cooking can be your place of inspiration, even if you are shrinking quickly. And who knows, your motivation does not find you even more motivated and productive... Free up time and even more to cook!

You are on your way to a healthier, more suitable, and food you eat (and how you eat it) is a central component to your success, so stick to it! As said, it goes... "If things don't change... Nothing changes! You are creating a difference, and by sticking to the simple tips mentioned above, you will make all your dream shift.

There is much content that has a lot of useful tips on how to eat healthily and stay in shape. You will come to understand why staying fit is essential for your body and health. This cookbook gives you good ideas about what type of diet you should operate every day and what kind of food and drinks they add.

You will also receive tips on how you can maintain your diet even when it's hard to keep. It makes anyone sometimes and mistakes, but there are ways to maintain a specific diet to improve your overall body health.

Some people are not fully aware that they need to work on their health. They eat whatever you want and assume that since they are in the ideal weight range they are healthy. This is not necessarily true. As in reading, you will learn about the right ways to keep your body healthy.

Nutrition Guidelines

Reversing to diet drugs, fad diets are simply horrible, junk food should be avoided, the diet is taken exclusively. With all the crazy diet reviews out there, do you take into account the basics of healthy eating? Get rid of the confusion when it comes to dieting, and use the basics of high-quality nutrition tips and healthy food pyramid as your scaffolding for healthy eating.

So how do we know what nutritious meals should look like? Read on...

Nutritional Instructions That Work

To get an attractive skinny body, you are going to do a little more than just exercise. Most people think that if they exercise, then they can eat whatever they want. Most people use completely bad anyway (ranting for another day) and then they are eating horrible foods because they are "deserved".

Crazy.

Now you can see why we are a global obesity epidemic, if people who are exercising do so, those who do not stand a chance.

So what should be changed?

People need education (which should be taught at school, again no longer dispossession), the more you know about a subject, the better you will be at it. Knowledge is power. So you want to be lean? Start learning to lean.

Here are some simple rules that have been designed for fat loss, health and having an attractive leaner body. All you have to do is follow all your meals followed by these instructions, and you will lose fat.

Drinking water or green tea: If you are trying to lose weight, why drink extra calories? Your new favorite drink is water. The water has zero calories, which is suitable for what we are trying to achieve.

His body is 80% water, so it makes sense to keep him finished.

If you want to go a little crazy and have more than just water, green tea is your choice. This is a favorite of people trying to lose weight. As the metabolism increases, I would like to mine the first thing in the morning, which is a great way to start the day.

Eating vegetables with every meal: vegetables are the best. As simple as that, if you do not follow for your best, then chances are you need to eat more. These are perfect to the edges well and are very difficult to overeat. You will get a lot of energy for all your daily activities of vegetables.

Carbohydrates are only allowed after exercise: most people eat too many carbs. Carbohydrate is the preferred energy source for the body and is rich in calories and therefore energy, the problem is that people use power to put them into their bodies. When this happens the body uses this extra energy and stores it just about as fat.

As a pet on the back for training and exercise, you can use some carbohydrates to return some of the energy you are used to. If you don't use excess energy during the day, you get carbs. It's a rewarding system and it works great.

Eat protein with every meal: a protein with every meal that will surely repair your body and we also need it to keep our muscle mass in our frames. As your active tissue muscles make a lot of sense to keep them as perfect as possible, that's why we train with weights, and protein building blocks have this muscle development and maintenance.

Eating healthy fats: Long days, we recommend staying away from fat. But many people still avoid fat like the plague. And it's not a good move, a little fat is essential and helps you lose body fat... Strange but real.

Good fats are called polyunsaturated fats and are found in avocado, olive oil, almonds, and sesame seeds.

Feed the missing link between good results and exceptional results and remember that you can't train a bad diet. I hope this has clarified a few things, but remember that this story is never over and you have a lot of time and effort researching and adjusting your diet for results.

Understanding Calories

No matter how many fad diets come with claims, allow you to eat whatever you want in any amount, if you violate the fundamentals of basal metabolism, you won't lose weight. Put, if you consume more calories than you spend through your body, you need natural and exercise, you'll lose weight. Ethical nutrition principles are an equal number of calories you'll consume with the number of calories you burn. If you consume more calories than burning your body, it will become extra calories for fat. If you burn over your intake, your body will burn additional calories needed from your fat store and lose weight. If your calories are consumed and burned in balance, your weight will be preserved.

The first step in understanding calories is to determine how many calories you need daily. The number of calories you need depends primarily on your activity level. The average man needs around 2,800 calories, while athletes can burn up to four or even 5000. There are three basic levels of activity in which most people fall:

- ☐ Sedentary: These levels are defined as low exercise or no exercise in the day. Men and this level require about 2,400 calories, and women need 2,000.
- ☐ Active: This level is defined as 30-60 minutes of exercise per day. Men and this level require about 2,800 calories, and women need 2,200.
- ☐ Very active: This level is defined as more than 60 minutes of exercise per day. Men and this level require about 3,000 calories, and women need 2,400.

To get a very accurate measure of your daily caloric needs, you can use a tool called Harris-Benedict Formula (HBF). HBF uses its weight, age, height, and sex to calculate its basal metabolic rate (BMR), which is defined as its number of calories if you sleep all day. Then, by multiplying your BMR with your activity level, you can get a more thorough study of your daily caloric needs. HBF uses the metric system to remember to convert pounds in kilograms and inches to centimeters. You can do this by multiplying 0.4536 pounds and inches by 2.54.

The first step in using the Harris-Benedict Formula is to calculate your BMR:

- For men: BMR= 66 + (13.7 x weight in kg) + (5x height in cm) - (6.8 x age in years)
- For women: BMR= 655 + (9.6 x weight in kg) + (1.8 x height in cm) - (4.7 x age in years)

Once you've calculated your BMR, multiply it by your activity level.

- Sedentary: BMR x 1.2
- Active: BMR x 1.55
- Very Active: BMR x 1.725

Once you've calculated how many calories you need, you can adjust your caloric intake to manage your weight. If you want to lose weight quickly, aim for 500 less calories a day. Pound fat is 3,500 calories so by creating a daily negative calorie balance of 500 calories you can burn a pound of fat per week. So this science is behind any weight loss program, but sticking to the challenge plan. The night of time can ruin the day of effort, so try a partner to get you in check and plan your daily meals.

Calorie Count and Macronutrients

A healthy eating illusion is a controversial and confusing subject. Whichever are their claims about what they consider nourishing? Diets that rely on calorie counts or macronutrients can be misleading and inaccurate. Not all calories are created equal, nor are sources of fat, carbohydrates, and proteins. The essential factor is the quality of food and ingredients used to produce food. The concept of counting macros and calories makes sense. Your body needs enough food to provide energy for physical demands imposed on it. The higher your level of activity, the more calories you need. The problem with this concept is that we look at the numbers on the labels and ingredients. By emphasizing how much protein, fat and calorie food is, the ingredients don't even consider.

The box grains are processed with preservatives and added sugars. Most sausages are made with nitrates, high sodium levels, and low-grade meats. Coincidentally, both samples are marketed as healthy.

Suppose you have a bowl of oatmeal prepared with sliced bananas and snapped cereals. Instead of sausages, you are part of grass-fed beef. Even if the calories are equivalent to both replacements, they are cleansing options. Take the time to find out what the food you are eating. The ingredients are essential. They are looking for foods with minimal processing. Organic foods have higher standards that must be fulfilled as such is certified. Food with "All natural" claims is not necessary to meet any benchmark for any manufacturer who claims.

Red food or drink may be attributed to crushed insects. Refer to Carmine and snail, which are used as distances in food, beverages, and cosmetics. Perhaps the wood particles blend with your cheese. Cellulose is composed of wood pulp and processed in a way that will not occur in nature. Maltodextrin is a sweetener that is typically derived from corn in the United States. Likely, corn is genetically corrected. This add-on is found in a wide variety of foods and is often overlooked, while the corn syrup is high fructose all bad press. There are many food replacements like this, which find your way to your body.

If you do not feel food after eating, then the food should not enter the body again. Food must be cured and nourish your body. You should not feel restless or dense after a meal, and you should feel revived. Change your approach to eating. Information about ingredients simplify your diet, listen to your body and feel better. There is no need to read the ingredients if you eat real foods. We are able to find dense nutrient foods without having them being added artificially in a lab. Changing your eating habits is a vital step in creating a healthier and more appropriate one.

Meal Prep Made Easy – Adopt These Cooking Techniques

Cooking is a great themed dining experience that includes many different and varied ideas. Some methods are simple, while others are quite simple.

Prepare as much as possible. The preparation work must be done before it is time to start cooking. You can save yourself a lot of hassle by getting all your preparation work early.

An essential preparation is to cook a meal for family or friends. Make sure you have all the ingredients ready. This can seriously reduce your stress levels and help manifest more positive results.

You're more pleased with this.

There are ways to repair your work and do it, so it doesn't get wasted. Mix 1 tablespoon of starch and 2 tablespoons of water. Add this mixture to your sauce while you make it simmer. Be sure to remove the seasoning and stop the starch slowly without stirring continuously to make it too thick.

Do you often throw a lot of mold fruits? Doesn't it cut the corrupt sector and save the rest? You should never eat or store a piece of fruit that begins to rot. Mold is much deeper than you see and can make you sick.

To create your crunch of fries, let the raw potatoes soak for about half an hour in cold water before frying.

Use fresh ingredients as they flavor more foods, and it can also be real protection money.

You should read the label when you are buying ingredients for a meal. Many typical cases contain unhealthy and unnecessary additives. You'll want to make sure it's not high in sugar or sodium.

Always make sure you are using any cooking oil you need to measure. This will help you reduce the fat now in your kitchen while cooking. It allows you to track the amount of oil you have with cooking.

Leave the bone on the barbecue when compressing for time. Meat cooks are faster this way because the bone pulls the heat into a pan center.

To avoid burning your food, it has been organized from the utmost importance to stay. Having all your cooking tools handy and organized will help you be the builder when it comes to cooking. If you are organized are, it's easy to lose clues to things and waste your money and make delicious food that you could eat!

Always follow instructions for cooking macaroni and cheese. The pasta will be perfect in this way and tasty because the good velvet cheese melts on top of the noodles. When serving the finished product, use a solid spoon. Season the dish with some types of peppers!

You can make your stocks. You have access to inventory in hand for soups and other containers. Creating your stock can reduce preservatives in the mixture that you have to avoid preservatives.

Let your cooked food sit a little before serving. Many home cooks don't realize how important it is to let the rest of the food. It's very tempting to serve meat as soon as you turn it off the grill. If you do this, you lose a lot of meat juice when you cut it. Always let your flesh to sit while.

The salt helps keep the plants on the board and add an extra flavor. Add additional salt while cutting plants. The salt used on this board gives your dish with plants a little extra flavor because it sticks to them.

There are better ways to make tortillas heat. You can use your oven at 350 degrees until they get crisp enough. You can also cook your omelet through the stove in the stove. Using one of these cooking techniques will result in a tastier final product.

This is when you tie Turkey with a baking rope. Cook your bird's sur evenly when you tie your feet and wings along your body so that cooking is done evenly. If you don't, the tips from them can be easily burned while the rest of the bird does not cook.

An excellent cooking method is to help maintain the moisture and taste in your chicken is first to brine the bird. Soak the chicken in saline water for at least an hour before cooking.

Set the clock timer parallel to you as well.

Countless food-infested meals have been making recipes harder than they need to. Sticking to simple and easy recipes is the best way to make sure you have a delicious and healthy meal in a short time. These tips can help you understand the cooking and make the taste of the food amazing.

Your Healthy Weekly Meal Plan

If you are like many people, you groan at the thought of doing meal prep. While you know it is something you need to do, it does not mean you like it! Meal prep takes time, but if you look at preparing say your vegetables just once for the whole week, then you will find it easier to eat healthy home-cooked meals each evening. Some foods, some vegetables are easy to prepare ahead of time and save well.

In this chapter, I have shared a sample meal prep plan you can follow to achieve your set goals.

Meal Prep Week 1

Monday

Breakfast: Beef and Bacon "Rice" with Pine Nuts
Launch: Quick & Easy Tomato and Herb Gigantes Beans
Dinner: Creamy Wild Mushroom One-Pot Gnocchi

Tuesday

Breakfast: Apple Pie Breakfast Farro
Launch: Chinese Sticky Wings
Dinner: Cauliflower Rice Deluxe

Wednesday

Breakfast: Chicken-Almond "Rice"
Launch: Pappardelle with Cavolo Nero & Walnut Sauce
Dinner: Black Bean & Avocado Tacos

Thursday

Breakfast: Raspberry Chia Breakfast Jars
Launch: Veggie Sausage & Sun Dried Tomato One Pot Pasta
Dinner: Spatchcocked or UnSpatchcocked Chicken with Vinegar Baste

Friday

Breakfast: Venetian "Rice"
Launch: Wicked Wings
Dinner: Pan-Barbecued Sea Bass

Saturday

Breakfast: Blackened Mexican Tofu, Greens, and Hash Browns
Launch: Roast Chicken with Balsamic Vinegar
Dinner: Company Dinner "Rice"

Sunday

Breakfast: Smoky Bean and Tempeh Patties
Launch: Orange-Five-Spice Roasted Chicken
Dinner: Soy and Ginger Pecans

Meal Prep Week 2

Monday

Breakfast: Nuts and Seeds Breakfast Cookies
Launch: Microwaved Fish and Asparagus with Tarragon Mustard Sauce
Dinner: Absolutely Classic Barbecued Chicken

Tuesday

Breakfast: Tilapia on a Nest of Vegetables
Launch: Orange-Tangerine Up-the- Butt Chicken
Dinner: Saffron "Rice"

Wednesday

Breakfast: Japanese Fried "Rice"
Launch: Almond-Stuffed Flounder Rolls with Orange Butter Sauce
Dinner: Hearty Quinoa Waffles

Thursday

Breakfast: Balsamic-Mustard Chicken
Launch: Sherry-Mustard-Soy Marinated Chicken

Dinner: Turkey-Parmesan Stuffed Mushrooms

Friday

Breakfast: Lonestar "Rice"
Launch: Two-Cheese Tuna-Stuffed Mushrooms
Dinner: Salmon Stuffed with Lime, Cilantro, Anaheim Peppers, and Scallions

Saturday

Breakfast: Blackened Mexican Tofu, Greens, and Hash Browns
Launch: Microwaved Fish and Asparagus with Tarragon Mustard Sauce
Dinner: Absolutely Classic Barbecued Chicken

Sunday

Breakfast: Smoky Bean and Tempeh Patties
Launch: Quick & Easy Tomato and Herb Gigantes Beans
Dinner: Creamy Wild Mushroom One-Pot Gnocchi

Meal Prep Week 3

Monday

Breakfast: Beef and Bacon "Rice" with Pine Nuts
Launch: Quick & Easy Tomato and Herb Gigantes Beans
Dinner: Creamy Wild Mushroom One-Pot Gnocchi

Tuesday

Breakfast: Apple Pie Breakfast Farro
Launch: Chinese Sticky Wings
Dinner: Cauliflower Rice Deluxe

Wednesday

Breakfast: Chicken-Almond "Rice"

Launch: Pappardelle with Cavolo Nero & Walnut Sauce
Dinner: Black Bean & Avocado Tacos

Thursday

Breakfast: Raspberry Chia Breakfast Jars
Launch: Veggie Sausage & Sun Dried Tomato One Pot Pasta
Dinner: Spatchcocked or UnSpatchcocked Chicken with Vinegar Baste

Friday

Breakfast: Venetian "Rice"
Launch: Wicked Wings
Dinner: Pan-Barbecued Sea Bass

Saturday

Breakfast: Blackened Mexican Tofu, Greens, and Hash Browns
Launch: Roast Chicken with Balsamic Vinegar
Dinner: Company Dinner "Rice"

Sunday

Breakfast: Smoky Bean and Tempeh Patties
Launch: Orange-Five-Spice Roasted Chicken
Dinner: Soy and Ginger Pecans

Meal Prep Week 4

Monday

Breakfast: Nuts and Seeds Breakfast Cookies
Launch: Microwaved Fish and Asparagus with Tarragon Mustard Sauce
Dinner: Absolutely Classic Barbecued Chicken

Tuesday
Breakfast: Tilapia on a Nest of Vegetables
Launch: Orange-Tangerine Up-the- Butt Chicken
Dinner: Saffron "Rice"

Wednesday
Breakfast: Japanese Fried "Rice"
Launch: Almond-Stuffed Flounder Rolls with Orange Butter Sauce
Dinner: Hearty Quinoa Waffles

Thursday
Breakfast: Balsamic-Mustard Chicken
Launch: Sherry-Mustard-Soy Marinated Chicken
Dinner: Turkey-Parmesan Stuffed Mushrooms

Friday
Breakfast: Lonestar "Rice"
Launch: Two-Cheese Tuna-Stuffed Mushrooms
Dinner: Salmon Stuffed with Lime, Cilantro, Anaheim Peppers, and Scallions

Saturday
Breakfast: Blackened Mexican Tofu, Greens, and Hash Browns
Launch: Microwaved Fish and Asparagus with Tarragon Mustard Sauce
Dinner: Absolutely Classic Barbecued Chicken

Sunday
Breakfast: Smoky Bean and Tempeh Patties
Launch: Quick & Easy Tomato and Herb Gigantes Beans

Dinner: Creamy Wild Mushroom One-Pot Gnocchi

Shopping Guide and Food List

Fruits

- ☐ Avocado (18 medium avocado)
- ☐ Banana- ripe (12 medium)
- ☐ Banana- unripegreen (1 medium)
- ☐ Banana-sugarlady finger (1 firm)
- ☐ Blueberries' (20 blueberries)
- ☐ Breadfruit (12 fruit)
- ☐ CantaloupeRock melon (12 cup)
- ☐ CarambolaStar Fruit
- ☐ Coconut (12 cup)
- ☐ CumquatsKumquats (4 pieces)
- ☐ Dragon fruit (1 medium)
- ☐ *Durian*
- ☐ Grapes, all types Guava- ripe
- ☐ Honeydew melon (12 cup)
- ☐ Kiwi fruit (2 small)
- ☐ Lemons & Limes (including juice)
- ☐ Longan (5 longans)
- ☐ Mandarin & Clementine
- ☐ *Oranges*
- ☐ Passionfruit (1 whole)
- ☐ Pawpaw
- ☐ Pineapple (1 cup)
- ☐ *Plantain*
- ☐ Pomegranate (14 cup seeds)
- ☐ Prickly pear
- ☐ Rambutan (2 rambutans)

- ☐ Raspberries (10 berries)
- ☐ *Rhubarb*
- ☐ Strawberries
- ☐ Tamarind (4 pieces)

Cereals & Grains

- ☐ Bran, Oats & Rice (2 tbsp.)
- ☐ Buckwheat groats (34 cup)
- ☐ Cereal, Gluten-free without honeydried fruit (1 cup)
- ☐ Flakes of corn (12 cup)
- ☐ Flakes of corn, gluten-free (1 cup)
- ☐ Flakes of quinoa (1 cup, uncooked)
- ☐ Millet (1 cup cooked)
- ☐ Noodles, rice stick & brown rice vermicelli (1 cup cooked)
- ☐ Noodles. soba (l3 cup)
- ☐ Oats (12 cup)
- ☐ Oats, quick (14 cup dry)
- ☐ Pasta (12 cup cooked)
- ☐ Pasta. Gluten-free (1 cup cooked)
- ☐ Polenta (1 cup cooked)
- ☐ Puffed amaranth (1 4 cup)
- ☐ Quinoa, all types (1 cup cooked) Rice, all types (1 cup cooked)

Flour

- ☐ Almond meal (14 cup)
- ☐ Buckwheat flour (23 cup)
- ☐ Cornmaize flourstarch (23 cup)
- ☐ Gluten-free flour (23 cup)

- ☐ Millet flour (23 cup)
- ☐ Potato flourstarch (23 cup)
- ☐ Qumoa flour (23 cup)
- ☐ Rice flour (23 cup)
- ☐ Sorghum flour (23 cup)
- ☐ Tapioca flourstarch (23 cup)
- ☐ Teff flour (23 cup)
- ☐ Yam flour (23 cup)

Nuts & Seeds

- ☐ Almonds, Brazil nuts, hazelnuts, pecans & walnuts (10 pieces)
- ☐ Chestnuts (20 boiled or 10 roasted)
- ☐ Flaxseedslinseeds (I tbsp.)
- ☐ Macadamias (20 nuts)
- ☐ Mixed nuts (20 nuts)
- ☐ Peanuts (32 nuts)
- ☐ Peanuts (1 tbsp.)
- ☐ Seeds- chia, egusi. poppy, pumpkin sesame (2 tbsp.)
- ☐ Seeds- sunflower (2 tsp)

Drinks

- ☐ Beer (1 can or 375ml)
- ☐ Drinking chocolate, cocoa, cacao but not carob (2 big tsp)
- ☐ Coconut water (12 cup or 100ml)
- ☐ Coffee- black & espresso (2 shots)
- ☐ Coffee- instant (2 tsp)
- ☐ Juice- cranberry & tomato (200ml)
- ☐ Juice- fresh orange (12 cup)

- ☐ Spirits- gin. vodka & whiskey (30ml)
- ☐ Strong Tea- green, peppermint & white (not dairy) (1 mug or 250ml)
- ☐ Weak Tea- black, chai & dandelion on water (1 mug or 250ml)
- ☐ Wine- red & white (1 glass or 150ml)

Herbs & Spices

- ☐ All herbs 8. spices, fresh & dried except garlic, onion or chicory (usually 1 tsp. check Monash app)
- ☐ Salt & Pepper
- ☐ Spice mixes (no garlic or onion)
- ☐ Stock without garlic or onion

Meat, Fish. Eggs, Tofu & Legumes

- ☐ Any unprocessed meat, fish or eggs (without high FODMAP ingredients like onion or garlic).
- ☐ Dahl- chana & urad (12 cup)
- ☐ Chickpeas, butter and garbanzo beans- canned & rinsed (14 cup)
- ☐ Lentils- canned & rinsed (12 cup)
- ☐ Lentils- red 8. green, boiled (14 cup)
- ☐ Lima & mung beans (14 cup)
- ☐ Quorum (75 g)
- ☐ Tempeh (100 g)
- ☐ Tofu- firm not silken (23 cup)

Vegetables

- ☐ Alfalfa (12 cup)
- ☐ Artichoke hearts, canned (18 cup)
- ☐ ArugulaRocket Asian & Collard greens

- ☐ AubergineEggplant (12 cup)
- ☐ 8amboo shoots Beans, green (12 beans)
- ☐ Beansprouts Beetroot (2 slices)
- ☐ Bell peppersCapsicum (12 cup)
- ☐ Broccoli heads or whole (1 cup)
- ☐ Broccoli, stalks or whole (12 cup)
- ☐ Brussels sprouts (2 sprouts)
- ☐ Cabbage (1 cup • not savoy)
- *☐ Carrots*
- ☐ Celery (5 cm stalk)
- ☐ Cetenac (12 medium piece)
- ☐ Champignons, canned (12 cup)
- ☐ ChardSilverbeet (1 cup)
- ☐ Chicory leaves (12 cup)
- ☐ ChicoryEndiveWitlof (4 leaves)
- ☐ Chili, red or green (11 cm long)
- ☐ Corn (12 cob max)
- ☐ CourgetteZucchini (12 cup)
- ☐ Cucumber (12 cup)
- ☐ Edamame beans (1 cup)
- ☐ Endive (4 leaves)
- ☐ Fennel bulb or leaves (12 cup)
- *☐ Galangal*
- *☐ Ginger*
- *☐ Kale*
- ☐ Leek leaves (12 cup)
- ☐ Lettuce and Endive- all types Mushrooms, oyster (1 cup)
- ☐ Okra (6 pods)
- ☐ Olives, green or black (15 small)

- ☐ *Parsnips*
- ☐ PicklesGherkins in vinegar (5 pieces)
- ☐ Potato- regular
- ☐ Potato- sweet potato (12 cup)
- ☐ PumpkinSquash- KentJapanese
- ☐ PumpkinSquash- butternut (12 cup)
- ☐ *Radish*
- ☐ Sauerkraut, white (1 tbsp.)
- ☐ Sauerkraut, red (12 cup)
- ☐ ScallionSpring onion (green tops)
- ☐ Seaweednori (2 sheets)
- ☐ Snow peasMangetout (5 pods)
- ☐ Spaghetti squash (1 cup)
- ☐ Spinach. baby (1 cup)
- ☐ Sprouts (12 cup)
- ☐ Spinach -English
- ☐ Tomatoes- regular
- ☐ Tomatoes, cherry (4 cherries)
- ☐ Tomatoes. Roma (I small)
- ☐ Tomatoes, sundried (2 pieces)
- ☐ Turnip, Swede, Rutabagas (1 cup)
- ☐ Water chestnuts (12 cup)
- ☐ Yam (1 cup)

Sauces & Condiments

- ☐ BBQ sauce (2 tbsp.)
- ☐ Capers (1 tbsp.)
- ☐ Chutney (1 tbsp.)
- ☐ AubergineEggplant dip (2 tbsp.)
- ☐ Mayonnaise (2 tbsp.)
- ☐ Mint sauce 8(Jelly (1 tbsp.)

- ☐ Miso paste (2 sachets)
- ☐ Mustard (1 tbsp.)
- ☐ Pesto sauce (12 tbsp.)
- ☐ Shrimp Paste (2 tsp)
- ☐ Soy. fish & oyster sauce (2 tbsp.)
- ☐ Sweet & Sour Sauce (2 tbsp.)
- ☐ Tahini (1 tbsp.)
- ☐ Tamarind paste (12 tbsp.)
- ☐ Tomatoes, canned (12 cup)
- ☐ Tomato sauce (2 sachets or 13g)
- ☐ Tomato paste (2 tbsp.)
- ☐ Vanilla essence (1 tbsp.)
- ☐ Vinegar- apple cider, malt, red wine, rice wine (2 tbsp.)
- ☐ Vinegar- balsamic (1 tbsp.) Wasabi (1 tsp)
- ☐ Worcestershire sauce (2 tbsp.)

100+ Healthy Meal Prep Recipes

Whether you are cooking for two or ten, or even more, you want your healthy meals recipes to be absolutely delicious. You also want them to be fast, easy and above all quick healthy dinner ideas. You do not want to rely on eating out frequently, nor do you want to be stuck preparing boxed, bagged, or frozen meals all of the time.

In this chapter, I have shared over hundred healthy meals recipes you can try out. They are quick and fast to prepare with an average time of 6 minutes.

Breakfast

Beef and Bacon "Rice" with Pine Nuts

Ingredients
½ head cauliflower
8 scallions, thinly sliced
23 cup (100 g) diced green pepper
1 tablespoon (28 g) butter
1 tablespoon (15 ml) olive oil
1 teaspoon chicken bouillon granules
¼ cup (60 ml) dry white wine
¼ cup (30 g) crumbled blue cheese
¼ cup (25 g) grated Parmesan cheese
2 tablespoons (30 ml) heavy cream

Instructions
Put cauliflower using a food processor with a razor blade. Put it in a microwave oven, add a tablespoon (30 ml) of water, and cover the microwave for 7 minutes over high heat.

While cooking, chop the onion and sauté the pepper. In a large, heavy fish, over medium heat, begin to marinate onions and peppers in butter and oil.

When the oven is "submerged", remove the cauliflower and drain it. When the green pepper begins to soften, dart the cauliflower at the fish and mix. Then mix in Bowl, white wine, blue cheese, parmesan and heavy cream. Cook and serve for another 3-4 minutes.

Yield: 5 servings

Each with 4 grams of carbohydrates and 1 gram of fiber, for a total of 3 grams of usable carbohydrates and 4 grams of protein

Apple Pie Breakfast Farro

Ingredients
8.8 ounces (249 g) quick-cooking dry farro
3 McIntosh apples or any favorite apple, cored and chopped
¼ cup (48 g) Sucanat or
1½ teaspoons ground cinnamon, plus optional extra for garnish
1 teaspoon pure vanilla extract
1 cup (235 ml) plain or vanilla vegan milk, warmed, as needed
1 or 2 recipes of nuts from Seed and Nut Ice Cream
Pure maple syrup, optional

Instructions
Boil a large pot of water. Add the farro and boil again. Lower the heat to medium-low and expose. Cook for 10 to 12 minutes until al dente or consistency is reached. Drain and reserve.

Place the chopped apples, lettuce or brown sugar and cinnamon in the same large pot used to cook the farro. Heat to medium-high, stirring to combine. Once the apples release their moisture, reduce the heat to medium and cook until the apples are tender, about 10 to 15 minutes, stirring constantly. Keep in mind that the cooking time will vary depending on the part of the apple and the type of apple you use. You're looking for tender bits, but not applesauce.

Drop the pot and mix the vanilla into the apples. Add the cooked grain into the apples and serve

immediately, topping each serving with as much of the warm milk as desired. Top each serving with a handful of nuts, extra cinnamon, and maple syrup if desired.
Yield: 4 to 6 servings
Protein content per serving: 19 g

Chicken-Almond "Rice"

Ingredients
½ head cauliflower
½ medium onion, chopped
2 tablespoons (28 g) butter, divided
1 tablespoon (6 g) chicken bouillon granules
1 teaspoon poultry seasoning
¼ cup (60 ml) dry white wine
¼ cup (30 g) sliced or slivered almonds

Instructions
Run the cauliflower using a food processor with a razor blade. Put the cauliflower in a microwave oven, a few tablespoons (30 ml) of water, cover and
Microwave for 7 minutes.
While cooking, heat the onion in a large butter (14 g) in a large skillet over medium heat.
When the cauliflower is over, remove it from the microwave, drain it and add it to the fish with the onion. Add the seeds, the chicken seasoning and the wine and mix. Reduce heat.
Let it boil for a minute or two and pour the almonds into a small, heavy fish, remaining in a tablespoon of butter. When the almonds are golden, pour them into the rice and serve.
Yield: 5 servings

Each with 4 grams of carbohydrates and 1 gram of fiber, for a total of 3 grams of usable carbohydrates and 2 grams of protein

Raspberry Chia Breakfast Jars

Ingredients
12 ounces (340 g) frozen raspberries, thawed but not drained
12 ounces (340 g) soft silken tofu or unsweetened plain vegan yogurt
¼ cup (80 g) pure maple syrup
2 tablespoons (24 g) maple sugar or (30 g) light brown sugar, optional
¼ cup (48 g) white chia seeds
¼ teaspoon pure vanilla extract
6 ounces (170 g) fresh berries (raspberries or blueberries), rinsed and thoroughly drained
Instructions
Place the melted raspberries in the blender or use a submerged blender to mix the berries. If you don't like blackberries, pass this mixture through a fine-mesh sieve. Add tofu or yogurt, maple syrup, and sugar to the berries and mix again to mix. Put in a large bowl.
Mix the chia seeds and vanilla in the mixture. Cover and refrigerate for at least 3 hours or overnight. Stir before serving.
Place some fresh berries in the bottom of the jar. (You can also mix the berries directly in the mixture and keep some to season).

Divide the chia preparation on top and sprinkle with the berries.

Remaining can be stored in an airtight container in the refrigerator for up to 4 days.

Yield: 6 servings

Protein content per serving: 7 g

Venetian "Rice"

Ingredients

½ head cauliflower

1 tablespoon (15 ml) olive oil

2 tablespoons (28 g) butter

1 cup (100 g) sliced mushrooms

3 anchovy fillets, minced

1 clove garlic, crushed

3 tablespoons (18.8 g) grated Parmesan cheese

Instructions

Run cauliflower through the food processor chopper. Put it in a microwave-safe bowl with a lid, add a few tablespoons (30 ml) of water, cover and simmer for 5 to 6 minutes.

When you're done, discover it right away! Combine olive oil and butter in a large, heavy skillet over medium heat and mix until well combined.

Add mushrooms and sauces to make the color soft and changeable. If the slices of mushrooms are large enough, you may want to break the edge of the spatula slightly when slicing.

When the mushrooms are soft, beat in the pan and stir. Add rice and brown without eggs - which will help a little water to mix. Mix well to distribute all the flavors.

Stir in the parmesan and serve.
Yield: 3 to 4 servings
Each will have 4 grams of protein. 2 grams of carbohydrates; 1 gram of dietary fiber; 1 gram of usable carbon.

Blackened Mexican Tofu, Greens, and Hash Browns

Ingredients
2 teaspoons onion powder
Preheat the oven to 300°F
2 teaspoons chipotle chili powder
1 teaspoon garlic powder
1 teaspoon smoked paprika
1 teaspoon dried oregano, crushed to a powder using fingers
½ teaspoon fine sea salt
1 pound (454 g) extra-firm tofu, drained, pressed, and cut into ½-inch (1.3 cm) slices
1 tablespoon (15 ml) high heat neutral-flavored oil
2 bunches (1½ pounds, or 681 g) Swiss chard, chopped
2 tablespoons (15 g) nutritional yeast
1 package (1 pound, or 454 g) hash browns
1 avocado, pitted, peeled, and sliced Salsa, for serving
Instructions
Combine the onion powder, chili powder, garlic powder, smoked paprika, oregano, and salt on a plate. Coat the tofu with the spice mixture. Heat the oil in a large skillet over high heat. Test the heat of the oil by dipping a corner of tofu into it. It should sizzle. Cook the tofu slices for 3 to 5 minutes until blackened. Turn

over to cook the second side for 3 to 4 minutes until also blackened. Keep warm in the oven.

Reduce the heat to medium. Put the Swiss chard into the same skillet. If the Swiss chard is freshly washed, it will still be slightly wet. If not. add a tablespoon (15 ml) of water, if needed, so it doesn't stick. Add the nutritional yeast and cook for 4 to 6 minutes, stirring occasionally, until wilted.

To serve, place one-quarter of the Swiss chard on each plate. Top with one-quarter of the hash browns and 2 to 3 pieces of tofu, depending on how many slices you were able to get. Place a few slices of avocado on the plate and serve the salsa on the side.

Yield: 4 servings

Protein Content Per Serving: 18 g

Smoky Bean and Tempeh Patties

Ingredients
1 cup (177 g) cooked cannellini beans
8 ounces (227 g) tempeh
½ cup (91 g) cooked bulgur
2 cloves garlic, pressed
1½ teaspoons onion powder
4 teaspoons (20 ml) liquid smoke
4 teaspoons (20 ml) Worcestershire sauce
1 teaspoon smoked paprika
2 tablespoons (30 g) organic ketchup
2 tablespoons (40 g) pure maple syrup

2 tablespoons (30 ml) neutral-flavored oil
3 tablespoons (45 ml) tamari
½ cup (60 g) chickpea flour
Nonstick cooking spray

Instructions

Mash the beans in a large bowl: It's okay if a few small pieces of beans are left. Crumble (do not mash) the tempeh into small pieces on top. Add the bulgur and garlic. In a medium bowl, whisk together the remaining ingredients, except the flour and cooking spray. Stir into the crumbled tempeh preparation. Add the flour and mix until well combined. Chill for 1 hour before shaping into patties.

Preheat the oven to 350°F (180°C*f* or gas mark 4). Line a baking sheet with parchment paper. Scoop out a packed ½ cup (96 g) per patty, shaping into an approximately 3-inch (8 cm) circle and flattening slightly on the prepared sheet. You should get eight 3.5-inch (9 cm) patties in all. Lightly coat the top of the patties with cooking spray. Bake for 15 minutes, carefully flip, lightly coat the top of the patties with cooking spray, and bake for another 15 minutes until lightly browned and firm.

Leftovers can be stored in an airtight container in the refrigerator for up to 4 days. The patties can also be frozen, tightly wrapped in foil, for up to 3 months.

If you don't eat all the patties at once, reheat the leftovers on low heat in a skillet lightly greased with

olive oil or cooking spray for about 5 minutes on each side until heated through.
Yield: 8 patties
Protein Content Per Patty: 10 g

Nuts and Seeds Breakfast Cookies

Ingredients
6 tablespoons (72 g) Sucanat
2 tablespoons (40 g) pure maple syrup
¼ cup (60 g) blended soft silken tofu or vanilla vegan yogurt
¼ cup (64 g) natural creamy cashew butter
2 tablespoons (30 ml) neutral-flavored oil
¼ teaspoon pure vanilla extract
Scant ¼ teaspoon fine sea salt
¼ teaspoon ginger powder or ground cinnamon
¼ cup (15 g) freeze-dried raspberries
3 tablespoons (30 g) shelled hemp seeds
1¼ cups (120 g) old-fashioned oats
¼ cup (90 g) whole wheat pastry flour
¼ teaspoon baking powder

Instructions
Preheat oven to 350°F (180°C, or gas mark 4). Line a large cookie sheet with parchment paper or a silicone baking mat.
In a large mixing bowl, combine the Sucanat, maple syrup, yogurt or tofu, cashew butter, oil, vanilla, salt, and ginger powder.
Add the berries, seeds, and oats on top. Sift the flour and baking powder on top.

Stir until well combined. Let stand for 5 minutes.

Scoop a packed ¼ cup (about 60 g) of dough per cookie onto the prepared sheet. Flatten slightly because the cookies won't spread a lot while baking. Repeat with the remaining 7 cookies.

Bake for 14 minutes or until the edges of the cookies are a light golden brown. Let cool on the sheet for 5 minutes before transferring to a cooling rack.

These are best served still warm from the oven or at room temperature. Store leftovers in an airtight container for up to 2 days.

Yield: 8 big cookies

Protein content per serving: 5 g

Japanese Fried "Rice"

Ingredients

½ head cauliflower, shredded

2 eggs

1 cup (75 g) snow pea pods, fresh

2 tablespoons (28 g) butter

½ cup (80 g) diced onion

2 tablespoons (16 g) shredded carrot

3 tablespoons (45 ml) soy sauce Salt and pepper

Instructions

With a microwave lid, add a few tablespoons (30 ml) of water, cover the microwave for 6 minutes.

While this is happening, beat the eggs, pour into a non-stick cooking pan and cook over medium-high heat. When you cook the eggs, use your spatulas to chop the size of the egg. Separate yourself from specialists and set aside.

Remove the peas and rows of snow peas and draw an inch (6 mm) long. (The microwave is already humming. Remove it from the cauliflower or turn it into a mushroom that doesn't look like rice at all!)

Melt the butter in a pan and leave the peas, onions and carrots for 2 to 3 minutes. Add cauliflower and mix well. Stir in the soy sauce and cook all the ingredients, stirring frequently for another 5-6 minutes. Add some salt and pepper and serve.

Yield: 5 servings

Each with 4 grams of protein; 5 grams of carbohydrates; 1 gram of dietary fiber; 4 grams of carbohydrates.

Tilapia on a Nest of Vegetables

Ingredients
1 pound (455 g) tilapia fillets
3 tablespoons (45 ml) olive oil
1 cup (150 g) red pepper, cut into thin strips
1 cup (150 g) yellow pepper, cut into thin strips
1½ cups (180 g) zucchini, cut in matchstick strips
1½ cups (180 g) yellow squash, cut in matchstick strips
1 cup (160 g) sweet red onion, thinly sliced
1 clove garlic, crushed
Salt and pepper
¼ teaspoon guar or xanthan
Lemon wedges (optional)
Instructions

Heat the olive oil in a heavy skillet over medium heat and leave the peppers, pumpkins, pumpkins, onions and garlic just 2 to 3 minutes and stir constantly.

Sprinkle the tilapia fillets with salt and pepper on both sides, then place the vegetables in the pan. Cover, heat slightly over medium heat and allow the fish to evaporate in the vegetable moisture for 10 minutes or until lightly crusted.

Using a slice, carefully transfer the fish to a serving plate and use a spoon to collect the vegetables over the fish. Pour liquid into the pan in the mixer and add guar or xanthan. Run the mixer for a few seconds and then pour the concentrated water over the fish and vegetables. To serve, place a slice of vegetables on a plate on each table and place a slice of fish on it. Some lemon wedges are right in this regard, but hardly necessary.

Yield: 4 servings

Each has 11 grams of carbohydrates and 2 grams of fiber, for a total of 8 grams of carbohydrates and 22 grams of protein.

Balsamic-Mustard Chicken

Ingredients

1 broiler-fryer chicken, about 3 pounds (1.4 kg), cut up, or whatever chicken parts you like

2 tablespoons (33 g) chili garlic paste

½ cup (120 ml) spicy brown mustard

¼ cup (60 ml) balsamic vinegar ¼ cup (60 ml) olive oil

Instructions

Put the chicken pieces in a large heavy plastic bag. Combine everything else and hit it together. Reserve a little marinade to lean on and then chop the rest with the chicken, squeeze the air and seal the bag. Pour the bag into the refrigerator and allow the chicken to marinate hourly all day.

When it's time to cook, turn on your coal or gas stove. You want a medium to medium fire. When the oven is ready, remove the chicken from the marinade using a tong and place it on a plate. Pour the marinade.

Now place the chicken on the grill skin and fry for 12 to 15 minutes. Turn it around and allow it to brown 7 to 9 minutes on one side of the skin. Turn it over and bake for another 5 to 10 minutes or read 180 degrees Fahrenheit (85 degrees Celsius) until the water clears when the bone is pierced and reads an instant thermometer. Repeat with the reserved marinade once again using clean dishes each time you eat. Keep the grill closed unless you cook or turn the chicken.

Yield: 5 to 6 servings

Assume 6 servings, if you have consumed all the marinade, each has 3 grams of carbohydrates and fiber, but you will get less. Assuming 6 servings, each will contain 30 grams of protein

Lonestar "Rice"

Ingredients
½ head cauliflower, shredded
1 tablespoon (15 ml) olive oil
1 tablespoon (14 g) butter
¼ cup (40 g) chopped onion

1 cup (100 g) sliced mushrooms
½ cup (40 g) snow pea pods, fresh, cut in ½ -inch (1.3-cm) pieces
¼ teaspoon chili powder
2 teaspoons beef bouillon granules or concentrate

Instructions

Put cauliflower with a lid in the microwave oven. Add 6 tablespoons (30 ml) of water, cover, and microwave for 6 minutes.

While cooking, heat the olive oil and butter in a large fish and smooth the onion, mushrooms and snow peas. I like to use the edge of the spatula to cut the mushrooms into smaller pieces, but if you want the way you cut them better - it's up to you.

When the mushrooms have changed and the snow pea is completely clear, drain the cooked rice and add it in the pepper and beef powder, mix to distribute the seasonings and then serve.

Yield: 3 servings

Each has 2 grams of protein; 5 grams of carbohydrates; 1 gram of dietary fiber; 4 grams of carbohydrates.

Lunch

Quick & Easy Tomato and Herb Gigantes Beans

Ingredients
2 tbsp. olive oil
1 onion
1 carrot

1 tsp ready-chopped garlic Protein content per serving garlic purée
1 Protein content per serving2 tsp paprika
400 g tin butterbeans
400 g tin chopped tomatoes
2 tbsp. tomato purée
1 tsp sugar
2 tsp dried oregano
handful baby spinach
handful fresh parsley
8-10 fresh mint leaves

Instructions

Heat the oil in a large pot or a large pot with oil. Chop onions and carrots and chop them finely and add them to the bowl with garlic and paprika. Cook over medium heat for 2 minutes.

Rinse and wash the potatoes and add to the pot, then add the greased tomatoes. Fill the empty tomato can in half with water and add it to the bowl with tomato puree, sugar, and oregano. Season well with salt and black pepper, boil, then reduce at dawn, cover and cook for 12-14 minutes.

Chop the baby spinach approximately, then add them to the pot and cook for 2 minutes. Chop the parsley and mint almost and stir just before serving. Taste and adjust the seasoning if necessary, then serve with crusty bread and crispy green salad.

Yield: 2 Servings

Chinese Sticky Wings

Ingredients

3 pounds (1.4 kg) chicken wings
¼ cup (60 ml) dry sherry
¼ cup (60 ml) soy sauce
¼ cup (60 ml) sugar-free imitation honey
1 tablespoon (6 g) grated ginger root
1 clove garlic
½ teaspoon chili garlic paste

Instructions

If complete, cut the wings to "resistance". Put the wings in a large plastic bag that can be used.

Mix everything else and reserve a little marinade to loosen and pour the rest into the bag. Seal the bag and press the air as you go. Turn the bag several times to cover the wings and chill in the refrigerator for several hours (a whole day is bright).

Preheat the oven to 375 degrees Fahrenheit (190 degrees Celsius or marked gas 5). Pull the bag, pour the marinade and place the wings on a shallow plate and allow them to cook for one hour in the oven, then marinate every 15 minutes. Use a clean container every time you eat.

Serve with lots of napkins!

Yield: Approximately 28 pieces

Each with 5 grams of protein, carbohydrate tracking; dietary fiber Tracking good carbohydrates Carbon numbers do not include honey-free mimicry.

Pappardelle with Cavolo Nero & Walnut Sauce

Yield: 2 Servings

Ingredients

200 g cavolo nero

150 g walnut pieces
250 g Pappardelle pasta linguine
1 slice bread
150 g dairy-free milk (soy, oat or nut milk)
2 tbsp. fresh parsley
optional - 2 tbsp. vegan parmesan or nutritional yeast flakes
1 clove garlic
olive oil

Instructions

Remove the kale stems and cut them into slices of protein per serving.

Heat a pan, peel the nuts (no oil needed) and bake over medium heat for 2-3 minutes. Turn off the heat and reserve.

Boil a large pot of water and soak Caverno Nero for 1 minute, then use a slotted spoon or tweezers to remove it and remove it with a sieve or stain (remove the boiling water in the pot).

Add the pappardelle to the boiling water and simmer for 8-10 minutes.

Meanwhile, sprinkle nuts, bread, milk protein content in each serving of milk, parsley, garlic, and Parmesan (if used) in a blender or food mixer and mix until consistent. Reach the thick sauce, beat it — season well with salt and black pepper.

Heat the pan again, add a little olive oil, and put it in Kawlow Norway. Cook for 3-4 minutes and turn off the heat.

When the pasta is cooked, drain it and return it to the pot. Tilt and add the walnut sauce to combine. Finally, add cavolo nero, overlay and then divide between two dishes.

Wicked Wings

4 pounds (1.8 kg) chicken wings
1 cup (100 g) grated Parmesan cheese
2 tablespoons (2.6 g) dried parsley
1 tablespoon (5.4 g) dried oregano
2 teaspoons paprika 1 teaspoon salt
½ teaspoon pepper ½ cup butter

Heat the oven to 350 degrees F (180 degrees Celsius or gas 4). Line a shallow pan with foil. (Do not miss this step or clean the pan a week later.)

Saving interesting things, cut the wings into "sticks". (Not sure what to do with wingtips? Freeze them for soup. Have a tasty broth.)

Combine parmesan and parsley, oregano, pepper, salt, and pepper in a bowl.

Melt the butter in a shallow bowl or pan

Soak each roast in butter, roll in cheese and spice mixture, and place in lined pan.

Bake for 1-hour B and then beat to avoid making a double recipe!

Yield: About 50 pieces

Each has only carbohydrates, fiber, and 4 grams of protein.

Veggie Sausage & Sun-Dried Tomato One Pot Pasta

Yield: 4 Servings
Ingredients
2 tbsp. olive or rapeseed oil
3 veggie sausages
1 onion, peeled and sliced
400 g pasta shells
200 g cherry tomatoes, halved
6-8 sun-dried tomatoes, roughly chopped
1-liter water
2 tsp vegetable stock powder
100 ml dairy-free cream (I used soya)
100 g fresh baby spinach
Instructions
Heat the oil in a large, shallow dish and fry the sausage and onion until the sausages brown. Carefully separate them from the pan and cut each piece into slices into 4 parts, then return to the pot for another 2 minutes.

Add pasta, tomatoes, sun-dried tomatoes, water and powder to the pot. Bring to a boil, then reduce to a sweet boil, cover and cook for 12-14 minutes, stirring every few minutes, until the pasta is well cooked.

Add the cream and spinach to the pot, then stir well and cook for another minute until the spinach has vanished.

Roast Chicken with Balsamic Vinegar

Ingredients
1 cut-up broiler-fryer Bay Leaves Salt or Vege-Sal Pepper
3 to 4 tablespoons (45 to 60 ml) olive oil
3 to 4 tablespoons (42 to 56 g) butter

½ cup (60 ml) dry white wine 3 tablespoons (45 ml) balsamic vinegar

Preheat the oven to 350°F (180°C, or gas mark 4).

Instructions

Wrap a sheet or two of leaves under the skin of each slice of chicken and sprinkle each slice with salt and pepper and place them in the grill pan.

Soak the chicken with olive oil and cover with the same butter. Roast in the oven for 1 V2 hour and rotate each piece every 20 to 30 minutes. (This makes the skin gloriously crisp and pleasant.)

When the chicken is ready, place it on a plate and pour the fat from the pan. Put the pan over medium heat and pour the wine and balsamic vinegar. Mix this loop and dissolve the delicious sweets that are glued in the cooking pan. Boil this in just a minute or two, pour into a pot or jar and serve with chicken. Throw in bay leaves before serving.

Yield: 4 servings

Each contains 2 grams of carbohydrates, fiber and 44 grams of protein

Orange-Five-Spice Roasted Chicken

Ingredients

3 pounds (1.4 kg) chicken thighs

¼ cup (60 ml) soy sauce

2 tablespoons (30 ml) canola or peanut oil

1 tablespoon (15 ml) lemon juice

1 tablespoon (15 ml) white wine vinegar

1 tablespoon (1.5 g) Splenda

2 tablespoons (40 g) low-sugar orange marmalade

2 teaspoons five-spice powder
Instructions
Put the chicken in a large plastic bag. Mix everything. Reserve a little marinade for weight loss and pour the rest into the container. Seal the bag and press the air as you go. Turn the bag over to cover the chicken and place it in the refrigerator. Allow for at least two hours, and this is a good time.

Heat the oven to 375 degrees Fahrenheit (190 degrees Celsius). Remove the chicken from the refrigerator, pour the marinade and place the chicken in a pan and fry the chicken for 1 hour.

Reserved with Marinade 2 or 3 times, be sure to use clean containers every time you eat to avoid cross-contamination.

Yield: 5 to 6 servings
Assuming each will have 32 grams of protein.
3 grams of carbohydrates; dietary fiber tracking 3 grams of available carbohydrates - and you're supposed to consume whole marinade.

Microwaved Fish and Asparagus with Tarragon Mustard Sauce

Ingredients
12 ounces (340 g) fish fillets— whiting, tilapia, sole, flounder, or any white fish
10 asparagus spears
2 tablespoons (30 g) sour cream
1 tablespoon (15 g) mayonnaise
¼ teaspoon dried tarragon
½ teaspoon Dijon or spicy brown mustard

Instructions

Draw the bottom of the asparagus spears and cut them naturally. Put the asparagus on a large glass plate, add 1 teaspoon (15 ml) of water and cover with a plate. Microwave for 3 minutes.

While the asparagus is in the microwave, mix sour, mayonnaise, tarragon, and mustard.

Remove the asparagus from the microwave oven, remove it from the pie plate, and set aside. Drain the water from the runway. Put the fish fillet in it

Peel the pie plate and spread 2 tablespoons (30 ml) cream mixture on them and cover the pie again and place the fish in the microwave for 3 to 4 minutes. Open the oven, remove the plate from the top of the pie plate and place the asparagus on top of the fish. Cover the pie plate again and cook for another 1-2 minutes.

Remove the pie plate from the microwave oven and remove the plate. Put the fish and asparagus on a serving platter. Chop any boiled sauce on a plate over fish and asparagus. Melt each with reserved sauce and serve.

Yield: 2 servings

Each with 4 grams of carbohydrates and 2 grams of fiber, for a total of 2 grams of usable carbohydrates and 33 grams of protein

It also packs 949 mg of potassium!

Orange-Tangerine Up-the- Butt Chicken

Ingredients

3½ to 4-pound (1.6 to 1.8 kg) whole roasting chicken

1 teaspoon salt or Vege-Sal
1 teaspoon Splenda
1 drop blackstrap molasses (It helps to keep your molasses in a squeeze bottle.)
1 teaspoon chili powder
3 tablespoons (60 g) low-sugar orange marmalade
1 12-ounce (360-ml) can tangerine Diet-Rite soda, divided (Make sure the can is clean!)
2 to 3 teaspoons oil
1 teaspoon spicy brown mustard

Instructions

Prepare your grill for indirect cooking - if you have a gas stove, light it on one side. If using charcoal, place the lighter on one side of the grill and light.

Remove the chicken's neck and towels and wash the chicken and dry it with paper towels.

In a small bowl of salt or Vege-Sal, mix Splenda, molasses, and red pepper powder. Pour half of the mixture (18 teaspoon) into a bowl and store. Rub the rest into the chicken hole.

Mix the orange marmalade with low sugar in the reserved spice mixture. Open the tangerine boxes and pour 3.2 cups (160 ml). Put N cup (60 ml) of the beverage in a jamspice mixture and mix - you can drain or throw away the remaining siphon you poured. Now, using a church-style console door, pull several holes at the top of the can. Cover the box with non-stick cooking spray and place it in a deep bowl. Carefully place the chicken on the tables and place the can inside the chicken cavity. Rub the chicken with oil.

Okay, you're ready to cook! Make sure you place a pan. Set the chicken to rotate vertically on the beverage cans on the side of the oven, not above the fire, and gently spread the drums to create a tripod effect. Close the grill and cook the chicken at 250 degrees Fahrenheit (130 degrees Celsius) or about 75 to 90 minutes or until the juices are evident when the bone sticks. You can also use a meat thermometer. Must have 180 degrees Fahrenheit (85 degrees Celsius).

While the chicken is cooking, add the mustard to the jam sodaspice mixture and mix all the ingredients. Use this mixture for the chicken dough for the last 20 minutes or fry.

When the chicken is finished, carefully separate it from the grill. Grill gloves are useful here or use hot pads and piles. Wrap the tins and separate them from the chicken and discard. Allow the chicken to rest for 5 minutes before sculpting. Meanwhile, heat each remaining sweetened sauce until boiling and serve as a chicken sauce.

Yield: 5 servings

Each serving contains 5 grams of carbohydrates and one trace of fiber. Assuming a chicken weighs 2.3 kilograms (1.6 kg), each meal will have 40 grams of protein.

Almond-Stuffed Flounder Rolls with Orange Butter Sauce

Ingredients

1 pound (455 g) flounder fillets, 4 ounces (115 g) each
4 tablespoons (56 g) butter, divided

2 tablespoons (30 ml) lemon juice
18 teaspoon orange extract
1 teaspoon Splenda
½ cup (45 g) almonds
¼ cup (40 g) minced onion
1 clove garlic, crushed
1½ teaspoons Dijon mustard
½ teaspoon soy sauce
¼ cup (15.2 g) minced fresh parsley, divided

Instructions

2 tablespoons (28 g) butter, lemon juice, orange juice, and splendor in a slow cooker. Cover the slow cooker, lower it and allow it to warm up as you secure the circular bearings.

Place the almonds in a food mixture and chop them into cornstarch. 1 tablespoon butter (14 g), melt the butter in a medium bowl and add the almonds. Bake almonds over medium heat for 5 to 7 minutes or until fragrant. Transfer them to a bowl.

Now melt the final spoon (14 g) of butter in a pan and boil the onion and garlic over medium-low heat until the onion is clear. Add them to the almonds and mix. Now mix the mustard, soy sauce and 2 teaspoons (7.6 g) of parsley.

Wrap the wool fillets on a large plate and divide the almond mixture between them and wrap it over the fillets, then glue them together and glue them with a toothpick.

Remove the lid from the slow cooker and mix the sauce. Put the rolls in the sauce and spoon the sauce over them. Cover the dish again and let it cook for 1 hour. When you are done, place the sauce on them and sprinkle the parsley left on it for serving.

Return: 4 servings

Each has 24 grams of protein, 5 grams of carbohydrates, 2 grams of dietary fiber, 3 grams of carbohydrates.

Sherry-Mustard-Soy Marinated Chicken

Ingredients

3 ½ to 4 pounds (1.6 to 1.8 kg) cut-up chicken
¼ cup (6 g) Splenda
3 tablespoons (45 ml) olive oil
3 tablespoons (45 ml) sherry
1 tablespoon (15 ml) mustard
1 tablespoon (15 ml) soy sauce
1 tablespoon (6.3 g) black pepper
½ tablespoon Worcestershire sauce
¼ cup (40 g) minced onion
1 clove garlic, crushed
2tablespoons (30 ml) water

Instructions

Mix everything, except the chicken, mix well. Put the chicken in a shallow pan, without reaction, or in a plastic bag. Reserve a little marinade to be weak and pour the remaining marinade over the chicken. If it is in a vessel, close it once or twice to cover it. If it is in a bag, remove the air, seal the bag and rotate it several times to cover the chicken. In any case, wrap the chicken in the refrigerator and let it boil for at least 1 or 2 hours and not be injured.

When the chicken is ready for the oven, place the stove on medium charcoal or charcoal covered with white ash. Rinse the chicken bone with the lid closed for 10 to 12 minutes (but check the flames from time to time!) And with the marinade reserved once or twice using a clean container, each time you cook, you lean. Turn the chicken loose and fry it on the skin for 6 to 7 minutes, then grate again, but now and then check to make it go away. Start the chicken in the bone, soak it again and grill with the lid closed for another 5 to 10 minutes until the water when the chicken water is cleaned off the bone when the chicken is pierced or until a thermometer reads 180 immediately. Delete it — Fahrenheit (85 ° C).

Yield: 5 to 6 servings

Suppose 6 and assume the consumption of whole marinade, each serving of 4 grams of carbohydrates and 1 gram of fiber, is 3 grams usable for the number of carbs. However, since you will not eat all the marinade, I count 2 grams per serving. 34 grams of protein.

Two-Cheese Tuna-Stuffed Mushrooms

Ingredients
½ pound (225 g) fresh mushrooms
1 can (6 ounces, or 170 g) tuna
½ cup (60 g) shredded smoked Gouda
2 tablespoons (12.5 g) grated Parmesan cheese
3 tablespoons (42 g) mayonnaise 1 scallion, finely minced

Instructions

Heat the oven to 350 degrees F (180 degrees Celsius or gas 4).

Clean the mushrooms with a damp cloth and remove the stems.

Barbecue tuna, gouda, parmesan, mayonnaise and onion

Mix well.

Pour the mixture into the mushroom covers and place in a shallow frying pan, add just enough water to cover the pan, cook for 15 minutes and serve hot.

Yield: approximately 15 servings

Each contains 1 gram of carbohydrates, one trace of fiber, and 4 grams of protein.

Dinner

Cauliflower Rice Deluxe

Ingredients

3 cups (500 g) cauliflower rice— about ½ head's worth

¼ cup (50 g) wild rice

¼ cup (180 ml) water

Instructions

Make your cauliflower rice as you want - Prepare the microwave oven, so be careful not to put it on the mushroom. You want to bid. Pour rice and wild water into a bowl, cover, and simmer until all the water is gone - at least half an hour, maybe a little longer. Cook cauliflower rice and mix in the wild rice and season as desired.

Yield: 8 servings

Even with wild rice, it has only 6 grams of carbohydrates, with 1 gram of fiber, for 5 grams of carbohydrates per cup (85 grams).

Creamy Wild Mushroom One-Pot Gnocchi

Yield: 2 Servings

Protein Content Per Servings: 13.5 g

Ingredients

5 small shallots

2 tbsp. rapeseed or olive oil

1 tsp ready-chopped garlic

250 g mixed mushrooms

500 g ready-made gnocchi

125 ml white wine

125 ml of vegetable stock

handful baby spinach

3 tbsp. soya cream

vegan parmesan-style 'cheese' (optional)

Instructions

Peel the shells and chop them finely. Soak the oil in a large pan or large saucepan and simmer for 2-3 minutes and soak the crumbs and garlic until soft.

Chop or halve the mushrooms (depending on their size) and add them to the pot. Turn on the heat and cook for 3-4 minutes to keep the water-free. Season well with salt and black pepper.

Add the gnocchi and then the white wine and let it bubble for a minute, then pour over the vegetables, return to a medium blender, cover with a lid and cook for 5 minutes, stirring occasionally.

Chop the spinach almost, then stir, followed by the soy cream. Adjust seasoning if necessary and adjust if needed, then serve immediately with a piece of vegan parmesan.

Black Bean & Avocado Tacos

Yield: 2 Servings

Ingredients

1 tbsp. rapeseed or sunflower oil
half a red onion
1 tsp ready-chopped garlic or garlic puree
200 g tinned black beans
5 cherry tomatoes
1 roasted red pepper
half an avocado
juice of half a lime
handful fresh coriander
2 tortilla wraps

Instructions

Heat the oil in a small pan. Peel and chop the onion and add to the pot and then the garlic. Cook for 2 minutes. Wash and wash black beans and put them inside. Mix well with salt and black pepper and cook over medium heat for another 3-4 minutes, stirring occasionally.

Meanwhile, cut the cherry tomatoes in half and chop the peppers into thin strips. Remove the stone from the avocado and cut it into 1 cm pieces. Combine the tomatoes, peppers and avocado in a small bowl and season with salt and black pepper. Press the lemon juice and pour everything into the mixture.

Heat a second pan and lightly brown the sides of the cream until lightly browned.

Chop the coriander leaves almost and mix them with the tomato and avocado mixture.

Place the roasted tortillas in layers with beans and onions, then mix the tomatoes, peppers and avocado and serve immediately.

Spatchcocked or UnSpatchcocked Chicken with Vinegar Baste

Ingredients

3 ½ pounds (1.6 kg) chicken— either whole or cut up
1 cup (240 ml) cider vinegar 3 teaspoons chili powder
2 tablespoons (3 g) Splenda
1 teaspoon cayenne 1 teaspoon paprika
1 teaspoon dry mustard 1 teaspoon black pepper ½ teaspoon cumin ½ teaspoon salt

Instructions

Technically, you'll have to cut both sides of the spine and obliterate it, but that's too much for me, so I'll leave my baby.

You look rough, I know it sounds hard, but if you have chicken or kitchen scissors - my Marta Stewart scissors are great at Kmart - it all takes about a minute and a half Along the bottom of the chicken, grab each one. Open the cut and the chicken and press on the bones to hear a slight gap. Now you have a flat chicken that you can place on the grill.

Or you just can't be bothered. Describe this process because it is very interesting, very trendy right now, and yet whole chickens are often very cheap. However, I think using chopped chicken is more comfortable, you know? These wastes work well with parts and do not have to be carved.

However, start to start the stove. You want it over medium heat. While the oven is warming, mix everything except the chicken.

Grill, the chicken that started from the skin and keep the grill closed, except those with frost or flame, for 15 minutes and each time using a vinegar mixture. Clean, liquid. Turn some of the skin down and fry for 7 to 9 minutes, still dull. Then pull the skin back on and continue to grill to clean the water when the chicken is pierced or record a thermometer reading 180 degrees Fahrenheit (85 degrees Celsius). Service.

Yield: 5 servings

Even with all the boiled liquid, you will only get 5 grams of carbohydrates, 1 gram of fiber or 4 grams of carbohydrate per serving - but you will not consume all the nutrients. Count over 2 grams per serving. 40 grams of protein

Pan-Barbecued Sea Bass

Ingredients
1 pound (455 g) sea bass fillets
1 tablespoon (8 g) Classic Barbecue Rub
4 slices bacon
2 tablespoons (30 ml) lemon juice
Instructions

Cut the sea bass fillets into pieces. Sprinkle both sides freely with a grill.

Grease a large, heavy fish with nonstick cooking spray and cook over medium-low heat. Using sharp kitchen scissors, pour the bacon into small pieces directly into a fisher. Shake it for a moment. As soon as you start cooking some fat from the bacon, wipe out a pair of fish and place the fish in the pan. Cover and set the oven timer for 4 minutes.

When the time is up, cook the fish and mix a little bacon to make it even. Restore the vessel and set the stopwatch for another 3-4 minutes. Look at your fish at least once. You don't want to overtake him!

When the fish has collapsed, you should serve the dishes and add the brown bacon. Pour lemon juice into the pan, mix and pour over the fish. Service.

Yield: 3 servings

Each with 31 grams of protein; 2 grams of carbohydrates; dietary fiber tracking 2 grams of carbohydrates.

Absolutely Classic Barbecued Chicken

Ingredients

3 pounds (1.4 kg) cut-up chicken (on the bone, skin on—choose light or dark meat, as you prefer)

1 ½ cup (40 g) Classic Barbecue Rub

½ cup (120 ml) chicken broth

½ cup (120 ml) oil

½ cup (120 ml) Kansas City Barbecue Sauce

Instructions

The chips or pieces of wood are softened for at least 30 minutes

Continue with the grill and set it for indirect smoking.

While the oven is warming, sprinkle the chicken with a spoon

Scrub Combine reserved scrub with broth and chicken oil to combine.

When the fire is ready, place the chicken on a greased pan, add wood chips or slices and close the grill. Let him smoke half an hour before you start smoking.

Then, each time you use a chip or more pieces, use a clean container each time you use it.

Smoke the chicken approximately 90 minutes or until an instant thermometer reads 180 degrees Fahrenheit (85 degrees Celsius).

When the chicken is finished, soak the skin in a Kansas City barbecue sauce and let it simmer for 5 minutes or on the fire near the skin.

Rinse the other side with the sauce using a clean bowl, wrap it, and put on the heat for another 5 minutes.

Boil the remaining sauce and serve with chicken

Avoid this same primary method using any scrub and any sauce!

Yield: 5 servings

For each serving of 9 grams of carbohydrates and 1 gram of fiber, there are 8 grams of available carbohydrates. 35 grams of protein.

Company Dinner "Rice"

Ingredients

1 small onion, chopped
1 stick (115 g) butter, melted
1 batch Cauliflower Rice Deluxe
6 strips bacon, cooked until crisp, and crumbled
¼ teaspoon salt or Vege-Sal
¼ teaspoon pepper
½ cup (50 g) grated Parmesan cheese

Instructions
Sprinkle the onion in butter until golden and brown. Cauliflower Delicious luxury cauliflower with onion and shaved bacon, salt, pepper, and cheese. Service.
Yield: 8 servings
Each with 8 grams of carbohydrates and 2 grams of fiber, for a total of 6 grams of available carbohydrates and 5 grams of protein.

Soy and Ginger Pecans

Ingredients
2 cups (200 g) shelled pecans
4 tablespoons (56 g) butter, melted
3 tablespoons (45 ml) soy sauce
1 teaspoon ground ginger
Instructions
Heat the oven to 300 degrees Fahrenheit (150 degrees Celsius or gas 2).
Sprinkle the muffins in a shallow frying pan and shake the butter and cover all the nuts.

Roast for 15 minutes, then remove from the oven and stir in the soy sauce. Sprinkle the ginger evenly over the walnut and mix.

Roast for another 10 minutes.

Yield: 8 servings

Each 6 grams of carbohydrates and 2 grams of fiber, for a total of 4 grams of usable carbohydrates and 3 grams of protein.

Hearty Quinoa Waffles

Yield: 6 to 8 waffles

Protein Content Per Waffle: 7 g

Ingredients

1½ cups (355 ml) water, divided

⅔ cup (119 g) chopped dates

3 tablespoons (42 g) solid coconut oil

3 tablespoons (60 g) pure maple syrup

1½ teaspoons pure vanilla extract

1¾ cups (210 g) whole wheat pastry flour

1 cup (185 g) packed cooked white quinoa

¼ cup (48 g) chia seeds

1 teaspoon baking powder

1 teaspoon ground cinnamon Generous

¼ teaspoon fine sea salt

Nonstick cooking spray

Instructions

Before starting, here's a quick note: It's best to make sure that all ingredients are at room temperature when making the batter so that the coconut oil doesn't solidify when combined.

Combine 1 cup (235 ml) of water and dates in a small saucepan. Bring to a boil, lower the heat, and cook on medium-high heat just until the dates start to fall apart; it should take about 2 to 3 minutes. Stir the coconut oil into the hot mixture to melt. Set aside to cool for at least 30 minutes. (Note that this can also be done in the microwave, using a deep, microwave-safe container and proceeding in 1-minute increments.)

Add the remaining 120 ml of cup of water, maple syrup and vanilla and stir.

Mix the flour, quinoa, chia seeds, baking powder, cinnamon and salt in a large bowl and stir to combine.

Pour the wet ingredients over dry to mix. Let stand according to the manufacturer's instructions while heating the waffle maker.

Waffle the iron easily with oil spray. 135 Add the waffle cup 135 cups (135 g) to two squares of waffle maker or follow the manufacturer's instructions enough to cover the dough so that it does not overflow and that the waffles are well cooked.

Close the waffle iron and bake for about 8 minutes until golden brown. Remove waffles from the iron and let stand on a cooling rack for at least 5 minutes until the waffles are crispy. Do not miss this step!

Leftovers are even better: you can toast them in a toaster or toaster so that the waffles are crispy again. You can also freeze waffles for up to 3 months, until you squeeze them. Still, freeze directly in the refrigerator or frozen toaster so they are hot and crispy.

Salmon Stuffed with Lime, Cilantro, Anaheim Peppers, and Scallions

Ingredients
1 whole salmon, cleaned and gutted, about 6 pounds (2.7 kg)
1 lime, sliced paper-thin
1 bunch cilantro, chopped
1 Anaheim chili pepper, cut in matchstick strips
3 scallions, sliced thinly lengthwise
2 tablespoons (30 g) olive oil
Instructions
Is simple. Heat the oven to 350 degrees Fahrenheit (180 degrees Celsius or gas 4). Put the salmon in a large saucepan sprinkled with non-stick cooking spray. Now pour everything except oil into salmon and distribute it evenly throughout the body cavity.
I want to use salmon
Hard needle and cooking cord. Rub it now with olive oil on both sides and bake 30 - 40 minutes. It is a good idea to stick a thermometer on the thick side of the meat to see if it is finished. It should read between 135 degrees Fahrenheit and 140 degrees Fahrenheit.
Slice slices with a few slices of spices per serving.
Yield: 12 servings
Each with 45 grams of protein; 1 gram of carbohydrates; dietary fiber following 1 gram of carbs.

Turkey-Parmesan Stuffed Mushrooms

Ingredients
1 pound (445 g) SKI

2 und turkey ¼ cup grated (75 g) Parmesan cheese
½ cup (115 g) mayonnaise 1 teaspoon dried oregano
1 teaspoon dried basil
2 cloves garlic, crushed
1 teaspoon salt or Vege-Sal
¼ teaspoon pepper
1½ pounds (670 g) mushrooms

Instructions

Heat the oven to 350 degrees F (180 degrees Celsius or gas 4).

Mix turkey, parmesan, mayonnaise, oregano, basil, garlic, salt and pepper and mix well.

Clean the mushrooms with a damp cloth and remove the stems.

Pour the mixture into the mushroom doors and place in a shallow bowl. Add just enough water to cover the pot, cook for 20 minutes and serve hot.

Yield: About 45 mushrooms

Each contains 1 gram of carbohydrates, one trace of fiber and 3 grams of protein.

Saffron "Rice"

Ingredients

½ head cauliflower
1 teaspoon saffron threads
¼ cup (60 ml) water
½ medium onion, chopped
1 teaspoon minced garlic or 2 cloves garlic, crushed
2 tablespoons (28 g) butter
2 teaspoons chicken bouillon granules
¼ cup (30 g) chopped toasted almonds

Instructions

Run the cauliflower using a food processor with a razor blade. Put the cauliflower in a microwave oven, add a few tablespoons (30 ml) of water, and cover the microwave for 7 minutes over high heat.

Start soaking the saffron wires in the water. While this is happening, pour the onion and garlic in medium heat butter into a large, heavy fish.

When the cauliflower is over, remove it from the microwave, drain it and add it to the fish. Pour in water and saffron and mix in the chicken seeds. Allow it to cook for a minute or two as you crush the almonds. Pour almonds into the rice and serve.

Yield: 5 servings of bright "rice" yellow

Each with 4 grams of carbohydrates and 1 gram of fiber, for a total of 3 grams of usable carbohydrates and 2 grams of protein.

Sides

Cauliflower Puree

Ingredients

1 head cauliflower or 18 pounds (680 g) frozen cauliflower

4 tablespoons (56 g) butter

Salt and pepper

Instructions

Cover the jar with a lid, a few tablespoons (30 ml) of water and cover. Boil it for 10 to 12 minutes or until chilled, but it doesn't smell. (You may want to steam or cook the cauliflower if you wish.) Empty it and place it in a mixer or pan to cook well. Add butter, salt and pepper to taste.

Yield: At least 6 generous portions

Each contains 5 grams of carbohydrates and 2 grams of fiber, for a total of 3 grams of usable carbohydrates and 2 grams of protein.

Chipotle-Cheese Fauxtatoes

Ingredients

1 large chipotle Chile canned in adobo, minced; reserve 1 teaspoon sauce

½ cup (60 g) shredded Monterey Jack cheese

1 batch The Ultimate Fauxtatoes

Instructions

Slice a teaspoon of Adobo sauce and place the chopped cheese in the final pepperoni bread. Serve immediately!

Each with 14 grams of protein; 14 grams of carbohydrates; 8 grams of dietary fiber; 6 grams of carbohydrates.

Cheddar-Barbecue Fauxtatoes

Ingredients

½ head cauliflower, cut into florets

½ cup (120 ml) water

½ cup (55 g) shredded cheddar cheese

2 teaspoons Classic Barbecue Rub or purchased barbecue rub

2 tablespoons (10 g) Ketatoes mix

Instructions

Insert cauliflower into your slow cooker, including stems. Cover the slow cooker, place it on a high surface and cook for 3 hours. (Or bake 5 to 6 hours.)

When it's time, use a spoon to remove the cauliflower from the slow cooker and place it in the mixer or food processor (hold the S blade in place) and place it in the Nymph there or you can pull the water out and use a hand mixer to bite cauliflower right in the pot. Let the cauliflower drain and clean it!

Mix everything to melt the cheese.

Yield: 3 servings

Each with 8 grams of protein, 6 grams of carbohydrates, 3 grams of dietary fiber, 3 grams of usable carbohydrates.

Hobo Packet

½ head cauliflower
½ medium onion
1 medium carrot
1 large rib celery
½ teaspoon salt
½ teaspoon pepper
8 slices bacon, cooked
2 tablespoons (28 g) butter

Instructions

Start the fire on the coal or heat a gas stove.

Chop the cauliflower into small pieces. Finely chop the onion, thicken the carrots one centimeter (6 mm) and chop the celery to a uniform thickness.

Pour an 18 cm (45 cm) heavy aluminum foil onto the countertop. Collect vegetables in the middle. Sprinkle with salt and pepper, crush
Bake the boiled bacon on top and fry it with butter. Bend the foil everywhere and sew several times to seal it well. Turn the end to close them.
Pour the whole package on the grill and cook for about 12-15 minutes over medium heat or on the stove. Bend the container with a knife, place it on a plate to open and serve.
Yield: 6 servings
Each serving contains 3 grams of carbohydrates and 1 gram of fiber, 2 grams for usable carbohydrates. 3 grams of protein.

Cauliflower Kugel

Ingredients
2 packages (10 ounces, or 280 g each) frozen cauliflower, thawed
1 medium onion, chopped
1 cup (225 g) cottage cheese
1 cup (120 g) shredded cheddar cheese
4 eggs
½ teaspoon salt or Vege-Sal
¼ teaspoon pepper Paprika
Instructions
Preheat oven to 350 ° F (180 ° C or gas mark 4).
Chop cauliflower into pieces of V2 (1.3 cm). Mix in a large bowl with onion, muffins, cheddar, eggs, salt and pepper and mix well.

Spray an 8-inch (20 20 20 cm) pan with nonstick cooking spray and spread the cauliflower mixture evenly downwards. Gently sprinkle the paprika on top and bake for 50 to 60 minutes or until the coagulant is set and light brown.

Yield: 9 servings of 2 grams of fiber, 3 grams of usable carbohydrates and 10 grams of protein.

Little Mama's Side Dish

This is just the thing with a simple dinner of broiled chops or a steak, and it's even good all by itself It's beautiful to look at, too, what with all those colors.

Ingredients

4 slices bacon

½ head cauliflower

½ green pepper

½ medium onion

¼ cup (30 g) sliced stuffed olives

Instructions

It is fried in a large, heavy, medium-medium heat. (First, give the fisherman a jar of non-stick cooking spray.)

Chop cauliflower into 1.3 cm V2 inch slices. Also, remove the stalk. Put the shredded cauliflower in a microwave-safe pot without a lid, without wasting it, add a spoon (30 ml) of water, and cover the microwave for 7 minutes.

Mix the bacon and then return to the plywood. Grease the peppers and onions. Already some fat is cooked from the bacon and begins to brown around the edges.

Add the pepper and onion to the pan. Strain until the onion is translucent and the peppers start to soften.

The caution must be taken until the confluence of events ceases. Add it to the pot without stirring - mix a little extra water to dissolve the bacon aroma from the bottom of the pot and pour it into the pan with olive oil, leave it to cook for another minute. cook. When stirred, then serve.

Yield: 4 or 5 servings

Assuming 5 servings, each contains 3 grams of carbohydrates and 1 gram of fiber, for a total of 2 grams of usable carbohydrates and 2 grams of protein.

Gratin of Cauliflower and Turnips

2 ½ cups (375 g) turnip slices
2 ½ cups (375 g) sliced cauliflower
1 ½ cups (360 ml) carb countdown dairy beverage
¼ cup (60 ml) heavy cream
¼ cup (90 g) blue cheese, crumbled
½ teaspoon pepper
½ teaspoon salt
1 teaspoon dried thyme Guar or xanthan (optional)
¼ cup (25 g) grated Parmesan cheese

Instructions

Preheat the oven to 375 degrees Fahrenheit (190 degrees Celsius or marked gas 5).

Combine the shuttle and the cauliflower into a bowl to make sure they are almost evenly aligned.

In a small saucepan over low heat, heat the dairy and heavy cream to warm up, add the blue cheese, pepper, salt and thyme. Stir until the cheese is melted. It is good to thicken this sauce with guar or xanthan.

Spray a pot with non-stick cooking spray. Put about one-third of cauliflower flowers and turnips in the pan and pour one-third of the sauce evenly over them and make two layers of vegetables and sauce. Sprinkle parmesan on top — Bake for 30 minutes.

Yield: 6 servings

Each with 7 grams of protein; 8 grams of carbohydrates; 3 grams of dietary fiber; 5 grams of available carbohydrates.

Mushrooms in Sherry Cream

This is rich and flavorful and best served with a simple roast or the like.

Ingredients

8 ounces (225 g) small, very fresh mushrooms

¼ cup (60 ml) dry sherry ¼ teaspoon salt or Vege-Sal, divided

½ cup (115 g) sour cream 1 clove garlic 18 teaspoon pepper

Instructions

Clean the mushrooms and remove the wooden ends from the stems.

Put the mushrooms in a small saucepan with sherry and sprinkle with 18 teaspoon salt.

Bring the pears to a boil, reduce the heat, cover the pan and allow the mushrooms to boil for only 3-4 minutes and shake the pan once or twice while cooking.

In another small skillet over low heat, mix 18 teaspoon of salt, cream, garlic and pepper. You want to heat the cream through it, but don't let it boil or separate.

When the mushrooms are finished, pour the liquid into a small bowl. After the cream has warmed, spoon it over the mushrooms and mix everything over medium heat. If it looks a little thick, add a teaspoon or two of the stored liquid.

Stir in mushrooms and cream for 2 to 3 minutes over low heat, again making sure the cream does not boil or serve.

Yield: 3 servings

Each with 4 grams of carbohydrates and 1 gram of fiber, for a total of 3 grams of usable carbohydrates and 2 grams of protein

Avocado Cream Portobellos

Ingredients
6 small portobello mushrooms
¼ cup (60 ml) olive oil 2 cloves garlic
1 tablespoon (4.2 g) dried thyme
2 dashes hot pepper sauce
1 small black avocado
3 tablespoons (45 g) sour cream
2 tablespoons (20 g) minced red onion
Salt
6 slices bacon

Instructions
Start the fire on the coal or heat a gas stove.

Separate the stems from your portable (save them for cutting and serve for the tortilla or serve over the roast!) And place the mushroom covers on a plate. Measure the olive oil and chop one of the garlic cloves into it. Then mix the thyme and chili sauce. Using a brush, cover the launch caps on both sides with a mixture of olive oil.

Next, cut the avocado, remove the pits and pour it into a small mixer bowl. Fry it with a fork. Mix the cream, onion, and other garlic cloves. Add salt to taste.

Now we have to do your bacon cooking. Put it on a microwave bacon holder or a glass plate and cook over high heat for 6 minutes (depending on your microwave power it may be slightly different).

Grill the mushrooms while your bacon cooks! Place them on a slow charcoal oil stove or more than a medium and small gas stove. Roast for about 7 minutes or until the oil mixture is constantly beaten - you also want to use a bottle of water to put out the flames.

When your mushrooms appear to be browning, place them back on your plate and extend them back into the kitchen. If it's not clear, give it another minute or two and then drain it. Divide the avocado mixture among the mushrooms, mixing it nicely and well. Chop and serve a slice of bacon on each stuffed mushroom.

Return: 6 servings

For every 9 grams of carbohydrates and 3 grams of fiber, there are 6 grams of usable carbohydrates. 6 grams of protein; 701

Grilled Portobellos

Ingredients
4 large portobello mushrooms
½ green pepper
¼ small onion
1 clove garlic, crushed
¼ cup (60 ml) olive oil
Salt and pepper
¼ cup (25 g) grated Parmesan cheese

Instructions
Start charcoal fire or preheat a gas grill.
Separate the stems from your portable (save them from cutting in the tortilla or draining to go on the stick) and place the lids on a plate.
Cut both the green pepper and the onion and place it in the food processor with the S blade. Add the garlic and pulse to coat everything relatively thoroughly. Add the olive oil and press again.
Place the portobello on the side of the grill on medium heat and brush with a little oil in the green pepper mixture - just put a brush in. Allow mushrooms to brown for 4-5 minutes. Turn them and paste the green pepper and onion mixture into the pan for another 4-5 minutes. Watch out for the flaming olive oil! Sprinkle on a plate, sprinkle with salt and pepper and sprinkle each mushroom with 1 teaspoon (6.3 g) of parmesan. Service.
Yield: 4 servings

There are 9 grams of carbohydrates and 2 grams of fiber per portion of 7 grams of usable carbohydrates. 6 grams of protein

Kolokythia Krokettes

Ingredients
3 medium zucchini, grated
1 teaspoon salt or Vege-Sal
3 eggs
1 cup (150 g) crumbled feta
1 teaspoon dried oregano
½ medium onion, finely diced
I ½ teaspoon pepper
3 tablespoons (15 g) soy powder or (32 g) rice protein powder
Butter

Instructions
Mix the grated pumpkin with salt in a bowl and allow to stand for an hour or more. Remove and drain the liquid.
Mix and combine egg, feta, oregano, onion, pepper and soy powder
Good.
Spray a heavy spoon of fish with thick cooking spray, add 1 teaspoon (14 g) of butter and melt over medium heat. Fry the pumpkin stick with a spoon and turn it once during cooking. Add more butter if needed and keep the croquettes warm. The trick to this is to allow

them to completely brown or tend to disappear before trying to rotate them. If a few separate, do not sweat it. The pieces will still have an incredible taste.

Yield: 6 servings

2 grams of fiber, 4 grams of usable carbohydrates and 8 grams of protein.

Improve the preparation time for the dish by rolling a pumpkin and onion through a food processor.

Salads

Autumn Salad

Ingredients
2 tablespoons (28 g) butter
½ cup (60 g) chopped walnuts
10 cups (200 g) loosely packed assorted greens (romaine, red leaf lettuce, and fresh spinach)
¼ sweet red onion, thinly sliced
¼ cup (60 ml) olive oil 2 teaspoons wine vinegar
2 teaspoons lemon juice
¼ teaspoon spicy brown or Dijon mustard
18 teaspoon salt
18 teaspoon pepper
½ ripe pear, chopped
1 ½ cup (40 g) crumbled blue cheese

Instructions

Melt the butter in a small heavy saucepan over medium heat. Add the walnuts and

Allow to fry in butter and mix for 5 minutes.

While the walnuts are roasting and make sure you keep them and do not burn them, wash and dry the

vegetables and place them in an onion salad pan. They are first sprinkled with oil. Then combine the vinegar, lemon juice, mustard, salt and pepper and add to the salad bowl. Discard to cover everything well.
Melt the salad with pears, warm walnuts, and crushed blue cheese. Service.
Yield: 4 generous portions
And 6 grams of fiber, for a total of 7 grams of usable carbohydrates and 10 grams of protein

Classic Spinach Salad

Ingredients
4 cups (80 g) fresh spinach
18 large, sweet red onion, thinly sliced
3 tablespoons (45 ml) oil
2 tablespoons (30 ml) apple cider vinegar
2 teaspoons tomato paste
1½ teaspoons Splenda ¼ small onion, grated
18 teaspoon dry mustard Salt and pepper
2 slices bacon, cooked until crisp, and crumbled 1 hard-boiled egg, chopped
Instructions
Wash and dry the spinach very well. Tears the bigger leaves. Combine the onion in a salad bowl.
In a separate bowl combine oil, vinegar, tomato paste, espresso, onion, mustard and salt and pepper to taste. Pour the mixture over the spinach, onion and pour.
Chop the salad with bacon and eggs. Service.
Yield: 2 generous portions
2 grams of fiber, 5 grams of usable carbohydrates and 2 grams of protein.

Spinach-Strawberry Salad

Ingredients
1 pound (455 g) bagged, prewashed baby spinach
1 batch Sweet Poppy Seed Vinaigrette
1 cup (170 g) sliced strawberries
3 tablespoons (25 g) slivered almonds, toasted
½ cup (60 g) crumbled feta cheese
Instructions
Put the baby spinach in a large salad bowl. Pour dressing and pour well. It is served and served with strawberries, almonds and feta.
Yield: 4 servings
Each with 8 grams of protein; 11 grams of carbohydrates; 5 grams of dietary fiber; 6 grams of carbohydrates.

Summer Treat Spinach Salad

2 pounds (910 g) raw spinach
1 ripe black avocado ¼ cantaloupe
½ cup (15 g) alfalfa sprouts
2 scallions, sliced
French vinaigrette
Instructions
Wash and dry the spinach very well. Tears the bigger leaves.
Slice avocado in half, cut into crust and crust and cut into slices.
Peel and crush the cantaloupe or use melons.

Add avocado and cantaloupe to spinach with cabbage and lucerne. Throw in Winagart just before serving.

Yield: 6 servings

Each with 11 grams of carbohydrates and 5 grams of fiber, for a total of 6 grams of usable carbohydrates and 5 grams of protein.

Dinner Salad Italiano

Ingredients

1 head romaine lettuce, washed, dried, and broken up

1 cup (70 g) sliced fresh mushrooms

½ cucumber, sliced

¼ sweet red onion, thinly sliced

½ pound (225 g) sliced salami, cut into strips

½ pound (225 g) sliced provolone, cut into strips Italian or vinaigrette dressing

2 ripe tomatoes, cut into wedges

Instructions

Make a large salad bowl of salad, mushrooms, cucumbers, onions, salami and prolones. Roast with Italian tomato sauce or finger, then add and serve the chopped tomatoes.

Yield: 3 servings

Each with 17 grams of carbohydrates and 6 grams of fiber, for a total of 11 grams of usable carbohydrates and 36 grams of protein

Chefs Salad

Ingredients

iceberg, red leaf, or any other favorite lettuce

¼ pound (115 g) deli turkey breast

¼ pound (115 g) deli ham
¼ pound (115 g) deli roast beef
¼ pound (115 g) Swiss cheese
1 green pepper, cut into strips or rings
½ sweet red onion, cut into rings 4 hard-boiled eggs, halved or quartered
2 ripe tomatoes, cut vertically into 8 wedges each
Salad dressing
Instructions
Serve good salad beds on 4 plates.
Sliced turkey, ham, beef, and Swiss cheese into strips. (However, getting relatively thick meat and cheese is good for that.) Do all this artistically on salad beds and season with peppers, onions, eggs, and tomatoes. Let each diner add their clothes.
Yield: 4 servings
Each contains 13 grams of carbohydrates and 4 grams of fiber, for a total of 9 grams of available carbohydrates and 37 grams of protein.

Vietnamese Salad

Ingredients
4 cups (80 g) m lettuce, broken up
4 cups (80 g) torn butter lettuce
3 scallions, sliced, including the crisp part of the green shoot
1 ruby red grapefruit
1 tablespoon (1.5 g) Splenda
3 tablespoons (45 ml) fish sauce
3 tablespoons (45 ml) lime juice
1½ teaspoons chili garlic paste

2 tablespoons (15 g) chopped peanuts
½ cup (32 g) chopped cilantro ½ cup (12.8 g) chopped fresh mint

Instructions

Wash and dry the salad, combine and then divide it into 4 salad plates.

Chop the onion and spread on the salad.

Half-cut the grapefruit in half and use a sharp knife to slice it to break it apart. Divide portions of grapefruit into salads.

Mix Splenda, fish sauce, lemon juice and red pepper paste. Pour equal amounts of dressing on each salad. Then chopped ground peanuts, chopped cantaloupe, and pepper and serve.

Yield: 4 servings

Each with 4 grams of protein; 14 grams of carbohydrates; 4 grams of dietary fiber; 10 grams of usable carbohydrates.

Cauliflower Avocado Salad

4 cups (600 g) cauliflower
1 black avocado, peeled and diced
½ green bell pepper, diced
8 kg la mat olives, pitted and chopped
4 scallions, thinly sliced, including the crisp part of the green shoot Sun-Dried

Instructions

Cut the cauliflower into 1.3 cm (12 inch) slices. Put it in a microwave bowl, add 1 teaspoon (15 ml) of water and cover. Microwave oven for a long time for 7 minutes.

When the cauliflower is ready, drain it and pour it into a mixing bowl and set aside. Everything else, including sun-dried tomatoes and basil and chopped eggplant. Serve warm on a bed of salad.

Yield: 6 servings

Each with 3 grams of protein; 11 grams' carbohydrates; 4 grams of dietary fiber; 7 grams of usable carbohydrates.

Sour Cream and Cuke Salad

Ingredients

1 green pepper

2 cucumbers, scrubbed but not peeled

½ large, sweet red onion

½ head cauliflower

2 teaspoons salt or Vege-Sal

1 cup (230 g) sour cream

2 tablespoons (30 ml) vinegar (Apple cider vinegar is best, but wine vinegar will do.)

2 rounded teaspoons dried dill weed

Instructions

Mushrooms, cucumbers, onions and cauliflower as small as possible

It can be. The cutting blade works well in the food processor and saves you time, but do it with a good, sharp knife.

Brush the vegetables thoroughly with salt and refrigerate for an hour or two.

In a separate bowl, mix the cream, vinegar and dill and mix well.

Remove the vegetables from the refrigerator, remove any water that has accumulated on the bottom of the pan and mix the cream mixture.

Yield: 10 servings

1 gram of fiber, 3 grams of usable carbohydrate and 1 gram of protein.

Crunchy Snow Pea Salad

Ingredients
2 cups (150 g) snow peas
4 slices bacon
1 ½ cup (50 g) roasted, salted cashews
1 cup (160 g) diced celery
1 cup (150 g) diced cauliflower
½ cup (120 ml) ranch salad dressing
½ cup (120 g) plain yogurt
1 teaspoon spicy brown mustard

Instructions

First, you want to close the end of the snow pea and pull the ropes hard. Cut them into 1.2 cm (12 inch) pieces. Put the pieces of snow peas in a microwave bowl, add a spoon (15 ml) or water and cover with a plastic plate or foil. Just microwave for 1 to 2 minutes, then remove and uncover to stop the preparation.

Put the bacon on the microwave bacon
Tooth or on a glass footplate, microwave on high surface for 4 minutes or until crispy, then drain.
While the bacon is cooking, chop it raw. Combine all vegetables, including snow peas, in a bowl. Combine the dressing on the farm, the yogurt and the mustard. Pour over vegetables. Chop in bacon, add jars and throw again. Cool before serving.
Yield: 4 to 5 servings
Assuming that 4 will each have 8 grams of carbohydrates and 2 grams of fiber, 6 grams for usable carbohydrates. 5 grams of protein.

Parmesan Bean Salad

Ingredients
1 pound (455 g) bag frozen, crosscut green beans
½ cup (80 g) minced red onion
¼ cup (60 ml) extra-virgin olive oil
5 tablespoons (75 ml) cider vinegar
½ teaspoon salt or Vege-Sal
½ teaspoon paprika ¼ teaspoon dried ginger
¼ cup (75 g) grated Parmesan cheese
Instructions
Boil or microwave green beans until transparent.
Allow the beans to cool slightly and then mix the onion, oil, vinegar, salt, pepper, ginger, and parmesan. Cool well and serve.
Yield: 4 servings
Each has 12 grams of carbohydrates and 4 grams of fiber, for a total of 8 grams of carbs and 9 grams of protein.

Fish and Seafood

The Simplest Fish

1 fillet (about 6 ounces, or 170 g) mild white fish
1 tablespoon (14 g) butter
1 tablespoon (3.8 g) minced fresh parsley
Wedge of lemon
Instructions
Melt the butter in a heavy foil pan. Add the fish fillets and beat carefully for 5 minutes on each side or until the fish is matte and lightly lightening.
Transfer to a serving platter over the chopped parsley and serve with a lemon wedge.
Yield: 1 serving
Effect of carbohydrates, without fiber and 31 grams of protein.

Ginger Mustard Fish

4 (6 ounces, or 175 g) fish fillets, such as tilapia, cod, or orange roughly
4 tablespoons (56 g) butter
2 teaspoons minced garlic or 4 cloves garlic, crushed
2 teaspoons grated ginger
2 teaspoons spicy brown or Dijon mustard
1 tablespoon (15 ml) water
Instructions
In a large, heavy skillet, start the fish in butter over medium heat. It should be 4-5 minutes on each side. Take the fish on a plate.
Add garlic, ginger, mustard and more

Pour the water into the pan and shake everything well. Repeat the fish inside, carefully rotating it once to make sure both sides are familiar with the sauce. Let it cook for another minute and then serve. Remove the fish sauce.

Yield: 4 servings

Each contains 1 gram of carbohydrates, one trace of fiber and 31 grams of protein.

Aioli Fish Bake

Ingredients

1 fillet (about 6 ounces, or 170 g) of mild, white fish

2 tablespoons (30 ml) Aioli

1 tablespoon (6.3 g) grated Parmesan cheese

Instructions

Preheat the oven to 350°F (180°C, or gas mark 4).

Spray a shallow baking pan (an ideal jelly roll) with a colorless cooking spray. Right on the baking tray, sprinkle a thick fillet with Aioli and sprinkle with a tablespoon of parmesan. Carefully rotate the fillet and spread Aioli and sprinkle the remaining parmesan. Bake for 20 minutes.

Yield: 1 serving

1 gram of carbohydrates, one trace of fiber and 32 grams of protein.

Chinese Steamed Fish

Ingredients

12 ounces (340 g) fish fillets
2 tablespoons (30 ml) dry sherry
1 tablespoon (15 ml) soy sauce
2 teaspoons grated ginger
½ teaspoon minced garlic or 1 clove garlic, crushed 1½ teaspoons toasted sesame oil 1 or 2 scallions, minced (optional)

Instructions

Wrap the fish fillets on a piece of heavy aluminum foil and rotate the edges of the foil to create a lip.
Mix sherry oil, soy sauce, ginger, garlic, and sesame oil. Close a shelf - a cake cooling rack works well - in a large container. Pour about one centimeter (6 mm) of water into the bottom of the tray and increase the heat. Put the foil with the fish on it. Carefully pour the sherry mixture over the fish. Cover the pan well.
Cook for 5 to 7 minutes or until the fish is slightly crispy. If desired, serve with onion rolled like a pot.
Yield: 2 servings
Each contains 2 grams of carbohydrates, fiber and 31 grams of protein. Each serving has only 195 calories!

Wine and Herb Tilapia Packets

Ingredients
1½ pounds (680 g) tilapia fillets, cut into 4 portions
4 tablespoons (56 g) butter, divided
½ cup (120 ml) dry white wine, divided
¼ cup (16 g) minced fresh herbs (chives, basil, oregano, thyme, or a combination of these), divided
Salt
Instructions

Preheat oven to 350 ° F (180 ° C or gas mark 4).
For each sheet, pour a piece of aluminum foil about 45 cm (18 cm). Place a tab in the center of the foil square and gently twist the edges of the foil. 1 teaspoon (14 g) butter, 2 teaspoons (30 ml wine), 1 teaspoon (4 g) chopped herbs and just a little salt.
Wrap the foil around the fish, rotate the edges in the middle and down so that the packaging cannot fall into the oven. Repeat all 4 portions.
Put the packages right on the stove rack - without pan and bake for 35 minutes.
Yield: 4 servings
Each with 2 grams of carbohydrates and 1 gram of fiber, for a total of 1 gram of carbs and 31 grams of protein

Broiled Marinated Whiting

Ingredients
6 whiting fillets
½ cup (120 ml) olive oil
3 tablespoons (45 ml) wine vinegar
1 tablespoon (15 ml) lemon juice
1 teaspoon Dijon mustard
1 clove garlic, crushed
½ teaspoon dried basil
¼ teaspoon salt
¼ teaspoon pepper
mustard, garlic, basil, salt, and pepper and mix well.

Instructions

Put the fillets in a large plastic bag and pour it into the oil mixture. In the refrigerator for a few hours and rotate the container from time to time.

Preheat the meat. Separate the marinated fish. Bend about 20 cm (20 cm) over the fire for 4-5 minutes on each side or cook on a stove.

While the fish is cooking, pour the remaining marinade into a saucepan, boil briefly and then serve as a sauce.

Yield: 3 servings

Each has a little over 1 gram of carbohydrates, fiber and 34 grams of protein.

If you are in a hurry or do not have all the ingredients to prepare this dish, use the Winnie Garret dressing (180 ml) instead.

Whiting with Mexican Flavors

Ingredients
4 whiting fillets
2 tablespoons (30 ml) lime juice, divided
¼ teaspoon chili powder
2 tablespoons (30 ml) oil
1 medium onion
2 tablespoons (30 ml) orange juice
½ teaspoon Splenda
¼ teaspoon ground cumin
¼ teaspoon dried oregano
1 tablespoon (15 ml) white wine vinegar
½ teaspoon hot pepper sauce Salt and pepper
Instructions

Put the white fillets on a plate and sprinkle with 1 teaspoon (15 ml) lemon juice and return to the pallet. Peel the fillets without the red pepper powder.

Heat the oil in a heavy skillet over medium heat. Add the white fillets. Bake for about 4 minutes on each side, turning carefully or until cooked through. Take out a plate and keep warm

Add the onion to the pan and turn it over medium-high heat. Bake onions for a few minutes, until they begin to color. Mix lemon juice, orange juice, coriander, cumin, oregano, vinegar, and hot pepper sauce. Cook them all for a minute or two. Season with salt and pepper. Pour onion over fish and serve.

Yield: 4 servings

Each with 5 grams of carbohydrates and 1 gram of fiber, for a total of 4 grams of carbohydrates and 17 grams of protein

Each serving has only 162 calories!

Lightning Source UK Ltd.
Milton Keynes UK
UKHW020636190822
407545UK00009B/651